International Arrangements for Nuclear Fuel Reprocessing

International Arrangements for Nuclear Fuel Reprocessing

edited by
Abram Chayes
W. Bennett Lewis

Ballinger Publishing Company • Cambridge, Massachusetts
A Subsidiary of J.B. Lippincott Company

This book is printed on recycled paper.

International Standard Book Number: 0-88410-052-9

Library of Congress Catalog Card Number: 76-52961

Printed in the United States of America

Library of Congress Cataloging in Publication Data
Main entry under title:

International arrangements for nuclear fuel reprocessing.

1. Reactor fuel reprocessing–International cooperation. 2. Atomic energy industries–Security measures. I. Chayes, Abram, 1922– II. Lewis, W. Bennett, 1908–

TK9360.I58 621.48'38 76-52961
ISBN 0-88410-052-9

Contents

List of Figures

List of Tables

Acknowledgments

The papers in this volume were prepared for a symposium on International Arrangements for the Nuclear Fuel Cycle sponsored by the Pugwash Conferences on Science and World Affairs and organized by the Canadian and United States Pugwash Groups. The symposium was held at the conference facilities of the Johnson Foundation at Wingspread, Racine, Wisconsin.

The sponsors and organizers gratefully acknowledge the support of the Johnson Foundation, Racine, Wisconsin, The Rockefeller Foundation, New York, New York, Mr. Cyrus Eaton through the Pugwash Park Commission, and the Canadian Department of Public Affairs, University Program, whose contributions made it possible to conduct the symposium.

Introduction

Nuclear energy and nuclear weapons have, from the beginning, been strongly interconnected. That was the message of the Acheson-Lilienthal report at the beginning of the Atomic Age. Its truth has been confirmed by three decades of experience since that time.

Because the basic physical processes underlying peaceful nuclear power production are the same as those for atom bombs, it has never been possible to seal off completely the peaceful from the military uses of the atom. To spread the knowledge of one was, in some degree, to spread the knowledge and capability for both. And since, from the beginning, there was little prospect that mankind would deny itself the benefits of peaceful nuclear energy, nuclear policy–both national and international–has always had to struggle with the resulting dilemma.

The first effort to resolve the problem, embodied in the Baruch Plan proposed by the United States immediately after World War II, was abortive. But in 1956 the Statute of the International Atomic Energy Agency (IAEA) became the first major postwar accord on which the United States and the Soviet Union found a basis for agreement. The IAEA was established to help spread the benefits of nuclear energy throughout the world, but also to be a pivotal element in controlling the potential evils.

A decade later, in 1968, the superpowers again reached agreement on the draft Treaty on the Non-Proliferation of Nuclear Weapons (referred to as the NPT, or Non-Proliferation Treaty) which, after revision in U.N. debate, was opened for signature. Today almost 100 countries, including most of the important nuclear-capable powers, are parties to that treaty. Those parties that do not now have nuclear weapons have forsworn any right or intention to acquire them. And those that are presently nuclear powers have agreed not to assist others in so becoming.

It is true that a number of nations, including France, China, and others of particular concern, have been unwilling to adhere to the NPT. Nevertheless, the treaty embodies a broad international consensus against the spread of nuclear weapons. So strong is this consensus that even states that will not become parties to the treaty have announced that they will behave as if they were. And since its entry into force, no nation, whether a party or not, has avowedly pursued the policy of acquiring nuclear weapons. The consensus is of inestimable importance, and forms the keystone of all international non-proliferation policies.

The NPT embodies other international policies with respect to the atom. In particular, Article IV provides:

> 1. Nothing in this Treaty shall be interpreted as affecting the inalienable right of all the Parties to the Treaty to develop research, production and use of nuclear energy for peaceful purposes without discrimination and in conformity with articles I and II of this Treaty.
>
> 2. All the Parties to the Treaty undertake to facilitate, and have the right to participate in, the fullest possible exchange of equipment, materials and scientific and technological information for the peaceful uses of nuclear energy. Parties to the Treaty in a position to do so shall also cooperate in contributing alone or together with other States or international organizations to the further development of the applications of nuclear energy for peaceful purposes, especially in the territories of non-nuclear-weapon States party to the Treaty, with due consideration for the needs of the developing areas of the world.

And it was only right and proper to affirm that countries that forswore nuclear weapons were not giving up their technological birthright, and that countries already versed in nuclear energy had a responsibility to see that its benefits were widely and fairly shared.

As the recognition has grown in the 1970s that the earth's resources of fossil fuel are approaching their limits, nuclear power has entered more sharply into the future energy plans of many countries. It is too soon to say whether or not this move will be permanent. The past few years have seen some slowdown in the number of new nuclear plants going on line and in orders for the future. It seems clear that the forecasts of the 1950s and 1960s, particularly with regard to the developing countries, were too exuberant. There are some second thoughts about the breeder reactor. And some public resistance to the whole nuclear enterprise has begun to manifest itself in the United States and other industrial countries as well.

Nevertheless, as of fall 1976 the world has some 80,000 megawatts of installed nuclear generating capacity. Nineteen countries have plants in service and another twenty-two have plants under construction or on order. Of the first group, three are developing countries; of the second, another ten. So even on the most "antinuclear" assumptions, nuclear energy is bound to make a significant contribution to the world's energy supply, at the very least for the next few decades.

Because of the linkage between nuclear energy and nuclear weapons already noted, every country that acquires a nuclear power reactor has taken a step, at least potentially, toward a nuclear weapons capability. In particular, the operation of nuclear power reactors contributes to the assembly of an indispensable base of trained scientific and technical manpower. Nevertheless, power reactors of the current generation—light-water reactors (LWR's), basically of U.S. design, and heavy-water reactors of Canadian design—do not by themselves pose a serious threat of weapons proliferation. Without extensive processing, the reactor fuel is not suitable for making bombs, and the conditions of reactor operation make withdrawal or diversion of fuel for these purposes very evident. In these circumstances the safeguards imposed by the IAEA under the NPT or bilateral arrangements with suppliers can be considered adequate to detect any attempted diversion.

The case is different with other elements of the reactor-fuel cycle—particularly uranium enrichment and spent-fuel reprocessing facilities. And it is different on both counts: the material produced—uranium, suitably enriched, or plutonium—can itself be used as bomb material, and the difficulties of safeguarding the plants against unauthorized withdrawals or diversions are very much greater.

For the present there is no significant problem with respect to enrichment. The principal current technology—gaseous diffusion—remains classified, and resort to it is hindered by the enormous capital costs and power requirements and by the high visibility of any attempt to employ it. Technological developments may erode these barriers in the not too distant future, but they have not done so as yet.

As to reprocessing, the basic technology has been in the public domain since the 1950s. The nuclear weapons states have all extracted plutonium suitable for weapons from uranium that has been irradiated in "production" reactors. A number of other states have small-scale, experimental reprocessing facilities, some of which, as in India, South Africa, and Israel, are not under IAEA safeguards. Indeed, the plutonium for the Indian atomic explosion of 1974 was produced in its own unsafeguarded reprocessing plant.

In 1976 a number of major export transactions involving sales of large-scale reprocessing plants to advanced developing countries were discussed or concluded. In every case these facilities were included as part of a multireactor program. The West German-Brazilian deal has been the subject of considerable public comment, and much of the detail of the transaction is well known. France has agreed to sell plutonium-extraction facilities to Pakistan. A similar sale to South Korea was cancelled under U.S. pressure. Iran is actively in the market.

For the reasons outlined above, the spread of reprocessing facilities under the control of countries not now possessing nuclear weapons has become a matter of international concern and controversy. On the one hand, these facilities sharply enhance the opportunities for weapons proliferation, although reprocessing facilities are not presently necessary to produce nuclear power. On the other hand, the provisions of the NPT affirm the right of all countries to access to nuclear technology, and there are at least plausible reasons—unconnected with weapons development—why a nation might wish to begin reprocessing in the near future.

A number of approaches toward the problem of the spread of large-scale reprocessing facilities under national control have been discussed in various national and international forums. One of these is the principal subject of the present symposium—placing reprocessing facilities under international or multinational control. Under arrangements suitably devised, it was thought that nations might gain the benefits of early experience with reprocessing while avoiding dangers of national control.

Exploration of this approach was mandated by the NPT Review Conference of 1975, and the IAEA has undertaken a major long-term study of the concept. A number of individual governments have expressed interest, including the United States.

In these circumstances, it seemed to the United States and Canadian Pugwash Groups that the discussion might benefit from a systematic and dispassionate analysis of the concept, of the kind that the Pugwash Movement has been able to provide for major arms control and disarmament issues in the past. With the approval of the International Pugwash Movement, the symposium was organized and held at Wingspread, Racine, Wisconsin, from May 24 to 27, 1976.

The symposium addressed itself to a rather narrow and technical subject. We were not examining the whole gamut of antiproliferation policies, or even a generalized program for containing the spread of reprocessing and enrichment facilities. The analysis was focused on a single concept, its advantages and disadvantages, and some of the problems that would attend any effort to bring it into being. Our hope was that this analysis could feed into and inform the broader international discussions now under way.

The papers in this volume reflect that orientation of the symposium. After the two opening chapters in Part I, which are designed to provide the general background and setting of the problem, the main portion of the book is devoted to questions relating to nuclear fuel reprocessing. Part II deals with a number of the technical aspects of reprocessing: economic considerations, safeguards, health and safety, physical security, operational problems, and waste management. In each case the method is to consider the particular issue first in the context of national facilities, and then to ask whether significant differences, favorable or unfavorable, arise in the multinational case.

Part III explores some of the problems involved in setting up what would be a very complicated international institution to operate a high-technology venture. Some of the relevant experience, both inside and out of the nuclear field, is examined, and a general analysis of the institutional requirements is attempted. Part IV reverts to the broader political setting that conditions all non-proliferation efforts.

The symposium did not attempt to reach any agreed conclusions about multinational reprocessing facilities. All were agreed that there is no point in encouraging the establishment of any large-scale reprocessing facilities–national or multinational–unless it is clear that reprocessing will be necessary or desirable for the nuclear energy economy. But, on this question of the need for reprocessing the members of the symposium were divided. Some saw a present or immediate future need, usually in the context of early development of the breeder reactor. Some saw reprocessing as inevitable, but thought its advent could and should be delayed 20 or 30 years. Others continued to entertain hopes that it might be possible to bypass reprocessing entirely, and with it the so-called "plutonium economy" with all its dangers.

There was a considerable measure of consensus on a number of more concrete issues:

1. It would be well to start any multinational nuclear-fuel-cycle enterprise rather modestly, with relatively few participants, all of which have some extrinsic basis of cooperation, and to set limited objectives, spent-fuel management, for example.
2. In a full-blown multinational reprocessing enterprise, the gamut of considerations–economic, health and safety, nonproliferation, and security –argue for co-location of related facilities (spent-fuel storage, fuel fabrication, and waste-management operations) at the same site as chemical reprocessing.
3. The call for safeguards assumes the existence of adversaries motivated to depart from normal procedures and for cooperation among the operators and participants to defend against irregularities. With skillful management the involvement of nationals from several countries can enhance the possibilities for this cooperation.
4. The success of any multinational effort will depend on the free and voluntary adherence of the parties to any agreement. Political pressures to force participation are likely to be ineffective and to generate suspicion and resistance.

More generally, all were agreed that there is no single action or program –certainly no technical or institutional "fix"–that can eliminate the dangers of nuclear weapons proliferation. A major element in any such effort must be, as our concluding author has termed it, "psychological denuclearization"–

the recognition, by nuclear and non-nuclear powers alike, as well as their political electorates, that, in the words of the U.N. Secretary General on the eve of the adoption of the NPT:

> . . . Since the end of the Second World War no nuclear weapons state has been able to derive any immediate military advantage from the possession of nuclear weapons, let alone use them to gain an easy victory. . . .
> [Indeed,] the possession of nuclear weapons does not necessarily prevent a decline in political influence.

For the rest, all politically feasible actions operate on the margins of the problem. But it is not impossible that the cumulative effect of marginal actions can decisively influence the course of events. To the extent that the international community has been able to inhibit the spread of nuclear weapons in the first three atomic decades (and the effort has not been wholly unsuccessful), some such process has been at work. It is in this light that the question of international arrangements for the nuclear fuel cycle should be approached.

Part I

General Considerations

Chapter 1

The Nuclear Fuel Cycle and Nuclear Proliferation

George W. Rathjens
Albert Carnesale

The overwhelming majority of nations do not now have nuclear weapons, and of these the overwhelming majority, either by adherence to the Nuclear Non-Proliferation Treaty (NPT) or otherwise, have disavowed any intention to do so. The significance of these facts should not be underrated. Nevertheless, five countries, in addition to the United States, have crossed the nuclear threshold and detonated nuclear explosives since the beginning of the nuclear age. And despite the existing consensus, it can be expected that pressures in the same direction will operate on a number of nations in the future. This paper examines the courses such a nation might pursue, and, insofar as these involve the acquisition of national reprocessing or enrichment facilities, the range of alternatives available for generating countervailing pressures.

There are four paths leading to a nuclear weapons capability:

1. To construct facilities designed specifically to produce those materials (namely, uranium highly enriched in the isotope 235, uranium-233, and plutonium) suitable for the manufacture of nuclear weapons and "peaceful nuclear explosives."
2. To seize existing weapons.
3. To divert clandestinely from safeguarded nuclear-fuel-cycle facilities materials suitable for weapons.
4. To use weapons-grade materials produced in nuclear-fuel-cycle facilities in the absence of, following withdrawal from, or in disregard of any international agreements prohibiting such action.

To the extent that there is concern about terrorists or other subnational groups, only the second option–seizure of weapons–is likely to be important in the near term. For the somewhat longer time frame the last two options–both of which make use of nuclear-fuel-cycle facilities–also are worrisome possibilities. As to the question of acquisition of weapons by nations, which is the focus of this paper, all paths appear to be relevant.

The importance of reducing the likelihood of nations going down the third or fourth path will depend very much on how likely it is that they could acquire weapons through the first two routes. Obviously, seizure of nuclear weapons is possible wherever such weapons exist, but seizure by one nation of another nation's weapons is an important possibility in relatively few cases–in South Korea, in some countries in Europe (such as Italy and West Germany), and, less plausibly, in cases of naval ships in port.

The first path–a straightforward weapons program–is not so easily dismissed. In the case of each of the first five nuclear powers, a nuclear weapons program came first, with nuclear power programs following; and India, too, chose the direct path to its peaceful nuclear explosion, albeit in the context of a program where the principal emphasis was on the development of nuclear power. The direct approach is likely to dominate in the future as well, notwithstanding the rapid growth in nuclear power. States that are determined to acquire weapons (or peaceful nuclear explosives) are likely to find that construction of facilities dedicated to that purpose will be a cheaper and quicker route than development of a full nuclear power program which can then be used as a basis for a weapons program.

Consider first the production of plutonium. Even in a state having a substantial nuclear power program, the building of plutonium-production reactors probably would be preferred to relying on power reactors as a source of plutonium for weapons. This is especially likely if safeguards are applied to the power reactors and the fuels passing through them; but even in the absence of safeguards on the fuel cycle, the production reactor may be preferred for economic and practical reasons. The primary consideration from which this follows is that plutonium of a composition preferred for weapons manufacture (that is, plutonium which consists primarily of Pu-239 and which contains relatively small amounts of the heavier isotopes Pu-240 and Pu-241) is produced by removing the fuel from the reactor after relatively little exposure to radiation. Such early removal is quite inconsistent with operations optimized for power generation. The case for building plutonium-production reactors is especially strong if available power reactors are of types that must be shut down for fuel removal. The light-water reactors that now dominate the nuclear power industry all fall within this category.* It appears that to produce plutonium for weapons there is little if any economic cost and, indeed, possibly

*Other types of power reactors, of which the Canadian CANDU is the outstanding example, can produce plutonium of a composition that is desirable for weapons with far less economic penalty and with far less interference with normal reactor operations.

an economic advantage in spending $50–100 million on a plutonium-production reactor rather than in using, and interfering with the normal operation of, a power reactor in which the capital investment may be typically between $500 million and $1 billion.

The situation is somewhat different when one looks at another essential component of a plutonium-based weapons program; namely, a facility for separating the plutonium from the other elements (uranium, other actinides, and fission products) in the spent fuel. A separation plant, which is part of a nuclear power industry, can also be used to separate plutonium for weapons. Strong incentives for building an additional plant for a weapons program would exist only if the power-related separation plant were under safeguards or were being used to capacity. A small plant to produce materials for weapons could be built at a cost of tens of millions of dollars, as compared with a cost of hundreds of millions for a plant (subject to stringent environmental protection standards) designed to handle large amounts of heavily irradiated material from a power program.

Heretofore, the plutonium route to nuclear weapons generally has been preferred to the enriched uranium route. This reflects both the very large investment required for a uranium enrichment plant employing gaseous diffusion technology and the fact that weapons based on enriched uranium require more fissionable material than do those using plutonium (about three times as much if the comparison is between uranium containing at least 90 percent U-235 and plutonium containing very little Pu-240 and Pu-241). The plutonium route would probably still be preferred by most would-be nuclear powers.

Recent and likely future developments in uranium enrichment technologies —especially the centrifugal, aerodynamic, and laser enrichment technologies —increase the relative attractiveness of an enriched uranium weapons program (although there is still the need to use larger amounts of material than in the case of plutonium weapons). At a cost of tens of millions of dollars, any of about a dozen non-nuclear powers could probably build a uranium enrichment plant employing centrifuge technology that would be adequate to meet the needs of a small weapons program. The Becker aerodynamic nozzle process is an alternative that perhaps would be less suitable for a small program but probably also less demanding in precision manufacturing capabilities. It is likely, but by no means certain, that one of the laser enrichment processes will have been proven technically and economically feasible by the 1980s and, if so, enrichment for a small weapons program will probably be even more readily within the reach of technically advanced states.

As in the case of reactors and plutonium separation plants, nations having enrichment facilities under safeguards may choose to build additional small and specialized plants to produce enriched uranium for weapons rather than abrogate or circumvent any safeguards agreements. However, the safeguards matter aside, building new enrichment facilities for weapons programs could be attractive because plants designed to provide fuel for light-water reactors normally enrich fuel to about 3 percent U-235, whereas enrichment to approximately 90 percent would be desired for weapons. The incentives for building a specialized plant to produce weapons-grade material would depend on the demand for reactor fuel and on the adaptability of any existing or planned enrichment plants to produce highly enriched uranium. Of the proven technologies, gaseous diffusion plants are the least adaptable and centrifuge plants the most. Laser systems almost certainly would be more adaptable still, if they work at all.

One reluctantly concludes that within a decade, obtaining the key ingredients for a weapons program is likely to be a relatively easy and inexpensive matter for quite a number of states. The necessary knowledge for going the plutonium route on a small scale is already in the public domain, as is much information about the Becker nozzle process. As for the other uranium enrichment processes, probably the best that can be expected is a slight delay in the diffusion of knowledge, the effect of which will be that the plants built by new entrants will be less efficient than those capitalizing on more experience.

The first of the four paths to a nuclear weapons capability—the construction of facilities designed specifically to meet weapons production needs—is likely to dominate the choices. The second path—seizure of others' weapons—is not likely to constitute a significant threat. Closing access to the third and fourth paths—the clandestine or overt use of nuclear-fuel-cycle facilities—may effectively remove a national government's option of obtaining a nuclear weapons capability without making an unambiguous decision to do so. For both domestic and international political reasons, this could raise the threshold of the decision to acquire such a capability, but it is not likely to prevent access to the capability itself.

If it is indeed this easy for would-be nuclear powers to build facilities to produce the fissionable material required for weapons, and perhaps even expedient if they have nuclear-power fuel-cycle facilities under their control, is there a case for the establishment of multinational fuel-cycle facilities or for other measures that will make it less likely that nations will acquire their own power-related facilities for uranium enrichment and plutonium reprocessing?

Such a case can of course be made, although it should be borne in mind that this approach is only one of many possible approaches to inhibiting the spread of nuclear weapons. Other energy-related nonproliferation measures include strengthening the safeguards and physical security efforts of the International Atomic Energy Agency (IAEA), restrictions on nuclear exports, and assurance of supply of energy resources. In addition, there are measures of a political-military nature: strengthening the NPT, a comprehensive ban on all nuclear tests, progress toward the control of existing nuclear arsenals, sanctions against new nuclear weapons states, nuclear free zones, no-use pledges, and security guarantees. Some of these latter measures relate directly to the question of incentives to acquire nuclear weapons. Unless there is success in dealing with this problem, the energy-related measures are likely to be largely irrelevant. If developments in the international setting lead nations to want weapons, they will get them—multinational facilities, supplier agreements, and safeguards notwithstanding.

It is conceivable that a decision to begin a weapons program could be marginal, and that whether or not a nation has a uranium-enrichment or a plutonium-separation plant could be the decisive factor. But such a decision is likely to depend more on geopolitical considerations than on the existence of specific facilities. To the extent that technical considerations are important, the decision to develop a weapons capability is likely to depend more upon the availability and influence of trained technical personnel than upon the prior existence of facilities. This proposition is strengthened by the fact that the facilities involved (that is, enrichment or reprocessing facilities, and possibly even reactors designed especially for the production of plutonium) could be acquired with resource commitments on the order of tens of millions of dollars. While such a commitment might appear large to some developing nations, it is small compared to the cost of nuclear power plants. And, of course, only nations that have nuclear power plants are of interest in this context, since only they can claim to have a peaceful application for an enrichment or separation plant. Included among the nations likely to have substantial nuclear power programs within a few years will be Brazil, Iran, South Africa, South Korea, and Taiwan.

While inducing nations not to acquire their own "sensitive" fuel-cycle facilities cannot preclude forever their development of nuclear weapons capabilities, it can lengthen the lead time between the decision to acquire such a capability and the achievement of it. The longer the lead-time, the greater the likelihood that the decision will be reversed in response to domestic or international political change.

How much time could be bought? Perhaps as much as five to ten years in some of the less developed countries, but probably not more than one to three years in countries likely to have substantial nuclear power programs. As an example, consider Taiwan, which apparently plans to have several thousand megawatts of installed nuclear electric generating capacity in 1985. If at that time Taiwan has its own enrichment or reprocessing capability, it could probably detonate a nuclear explosion within three years of a decision to do so. If, on the other hand, the Taiwanese do not have their own sensitive facilities at the time of the decision, it is doubtful that the lead time to detonation would be lengthened by much more than one year, especially if Taiwanese technical personnel are involved in the design, construction, and operation of a multinational reprocessing facility.

On the assumption that the inducement of nations to forego the acquisition of independent sensitive facilities constitutes a worthwhile contribution to the nonproliferation effort, it is useful to examine mechanisms for bringing about such forebearance, recognizing that most, perhaps all, of the costs of implementing such mechanisms would have to be borne by those nations that place a high value on nonproliferation. Consider first the matter of enrichment facilities.

All nations that elect to build or buy light-water reactors (LWR's) as a component of power programs will require a source of uranium enriched from the natural isotopic abundance of 0.71 percent U-235 to approximately 3 percent U-235. Some other types of power reactors, none of which is expected to be of commercial significance in the twentieth century, require higher degrees of enrichment. (A high-temperature gas-cooled reactor, for example, would employ weapons-grade, highly enriched uranium as fuel.) Other kinds of power reactors, of which the Canadian CANDU is the most important example, employ natural uranium as fuel. Because approximately 90 percent of the power reactors in the world are LWR's, and this is likely to continue for some years, most nations with an interest in nuclear power will have an interest in an assured supply of enriched uranium.

For a number of years it has been possible for America's allies and purchasers of American reactors to secure enrichment services from the United States, which has developed excess capacity as part of its weapons programs. Enriched fuel for Soviet and Eastern European reactors has been supplied by the Soviet Union. If, however, power programs expand worldwide at the rates projected in the early 1970s, it is possible that a world shortage in enrichment capacity will develop by the late 1980s. The uncertainty in a prediction of enrichment demand is great, and is contributed to by uncertainties about the future of nuclear power and the economics of enrichment. Investment in a new enrichment facility is not an attractive proposition when there is a chance, as there is in the United States, of a slowdown in nuclear power plant construction. Nor is its attractiveness enhanced when enrichment technology is developing so rapidly as to threaten to make any new plant economically noncompetitive within a relatively short period of time.

To an individual entrepreneur considering an investment in a new enrichment plant based on gaseous diffusion or centrifuge technology, the threat of being undercut shortly by the arrival on the scene of a successful laser enrichment process looms quite large. But the investment required for enrichment capacity corresponds to only 3 to 5 percent of the cost of the nuclear plant network it serves. So from the point of view of a nation—or for that matter of an electric utility with a substantial nuclear power base—the economic risk will appear small compared to the benefit of an assured source of needed enrichment service. A decision whether to get into the enrichment business is likely to be based much more on the need for an assurance of fuel supply than on cost.

This says much about the kind of leverage the international community can use to induce nations to forego building such facilities. Assurance of supply will be important; cost considerations are not likely to be. It follows that the argument that enrichment services can be provided at lowest cost by large multinational (or American) facilities (with attendant economies of scale), and that also might employ a more advanced technology, is not likely by itself to dissuade a nation from acquiring its own facility. Even providing enrichment at a heavily subsidized price is not likely to dissuade the would-be builder concerned primarily with the matter of continuity of supply. It is worth noting that in recent years these inducements have grown weaker, and they can be expected to grow weaker still. There are several important reasons for this trend: (1) the ratio of enrichment costs to total fuel-cycle costs have dropped and can be expected to drop further; (2) with changing technology and the increase in the cost of the power required to operate enrichment plants, the economies of scale in enrichment have diminished, and can be expected to continue to do so; and (3) nations are likely to be more skeptical than in the past about sanctity of trade agreements in light of the trend toward autarky in trade relations (as evidenced by the OPEC oil embargo and oil pricing; by the United States reneging on agreements relating to exports of soybeans to Japan; by the unilateral reclassification of some of the Atomic Energy Commission's enrichment contracts as "conditional"; and by Westinghouse's unilateral cancellation of its contractual commitments to supply uranium).

If it seemed likely that there would be a surplus rather than a shortage in world enrichment capacity a few years hence, much of the concern about the possibility of an interruption of supply would be alleviated. This would be particularly so if there were a number of sources, for in that case buyers would be less concerned about possible interruption for political reasons. For assurance of supply, then, a buyers' market is desirable. This is within the power of the United States and the Soviet Union to bring about. There would be some cost to them, of course, but it would be small compared to the total investment in nuclear power.

The large multinational plant is another instrument for providing for enrichment while avoiding proliferation of national facilities. The effectiveness of such facilities in reducing concern about interruption of supply, particularly for the smaller participating nations, would depend on the mechanisms for allocating output and on the governing arrangements. A multinational approach is being implemented to some extent in arrangements under Eurodiff for joint funding of new gaseous diffusion plants in France. Iran, for example, has a significant minority share in the investment, carrying with it the right to a comparable proportion of the output, but no participation in management. It remains to be seen whether this will be regarded as a satisfactory arrangement by all of the participants. In particular it will be interesting to see if Iran will be willing to forego getting its own uranium enrichment capability *in Iran*.

The effectiveness of the establishment of a buyers' market in enrichment services or the establishment of multinational facilities to dissuade nations from acquiring their own indigenous capabilities would be greatly enhanced if dependent nations could maintain stockpiles of (or otherwise be assured of access to) enriched uranium sufficient to keep their reactors fueled for the several years they would require to build their own facilities should they be cut off from supply. Maintenance of such stockpiles would involve no risk for weapons proliferation, assuming that the fuel was enriched to only about 3 percent, as is sufficient for LWR's. But there would be a substantial economic cost involved in maintaining such stockpiles. If it is assumed that five-years' supply would be adequate, the carrying charges would be about $10 per kilowatt of nuclear electric power at current prices and an 8 percent interest rate.* Those requiring enrichment services should not be expected to pay such extra costs willingly. Perhaps they might be paid through direct subsidies or other arrangements made by the international community. Nor is there reason to believe that in the absence of action at governmental levels (that is, government construction of plants or guarantees for private investors), the required capacity will be developed in the form of multinational facilities or expanded programs in "safe" countries. Notwithstanding these difficulties, there are possibilities, in principle at least, of meeting the requirement for assurance of enrichment services as well as it could be met by indigenous capabilities.

Is there a way of going further, of making dependence on a multinational facility or on established suppliers more attractive than the building of indigenous facilities in terms of assurance of supply? There may be, at least for some countries. Because uranium ore is unevenly distributed among nations, most (but not all) countries concerned about the security of supply of enrichment services will also be concerned about the availability of uranium at reasonable prices. This means that, for most countries, arrangements guaranteeing supply of uranium as well as enrichment services could be very attractive. Thus, if nations with large uranium reserves would make long-term commitments to provide uranium to multinational facilities and to other established suppliers of enrichment services while refusing to export to others, there would be a strong disincentive for most nations to build their own enrichment plants. (Such an approach would not be especially useful for dissuading Australia, Canada, South Africa, or Sweden, all of which are believed to have large reserves of uranium that can be exploited at relatively low cost.)

It appears, then, that the proliferation of uranium-enrichment facilities can be prevented, or at least very much limited, by offering to those who will require enrichment services alternatives that meet the need for assurance of supply as well as, or better than, the construction of indigenous plants. The

*A small part of this could be offset by having capitalized on economies of scale that would presumably be inherent in enrichment at a multinational or other very large facility. On the other hand, there could be a further cost involved–probably less than $1 million –attributable to having produced enrichment five years earlier than needed in a situation where costs could be dropping rapidly (in constant dollars) because of technological change.

conditions can be met with either multinational facilities or expansion of plants in "safe" countries, by subsidizing the maintenance of stockpiles of enriched uranium, or by guaranteeing supplies of uranium to "approved" enrichment facilities. Presumably, any differential in the costs involved will have to be paid by those interested in holding the line on proliferation rather than by those to be deterred from acquiring indigenous capabilities. Unless this price is paid, and fairly promptly, there may well be widespread proliferation of enrichment facilities. The technical and economic impediments that have been highly relevant in the past will be far less so in the 1980s, and it is likely that during the last ten to fifteen years of the century most of the countries dependent on nuclear power will find enrichment technology within their reach.

Before turning from the enrichment issue, it should be noted that nations not already committed to LWR's have a realistic alternative not requiring enrichment—the CANDU reactor. Any attempt to deny nations an indigenous enrichment capacity in the absence of acceptable provision for security of supply from other sources will make the CANDU more attractive relative to the LWR. The commercial implications are obvious; the proliferation implications less so. Although CANDU enjoys a nonproliferation advantage over the LWR in not requiring enriched fuel, this advantage is offset—probably is more than offset—by its superior capability to produce plutonium for use in weapons manufacture. In contrast to LWR's, CANDU's are continuously fueled and thus could more easily operate with lower exposure of some fuel to radiation, producing, thereby, spent fuel containing plutonium characterized by a higher concentration of Pu-239.

Notwithstanding developments in uranium enrichment technology, for the next few years at least the plutonium route to a weapons capability will probably still be the preferred choice for most nations. We are thus concerned with the "back end" of the fuel cycle and particularly with the proliferation of capabilities for extracting plutonium from irradiated fuel.

In contrast to the situation with respect to enrichment, plutonium separation and recycling is not necessary to the operation of any current power reactors, whether they be LWR's, CANDU's, or any of the other kinds of reactors that provide the bulk of the world's nuclear power.* However, if the spent fuel were reprocessed, the separated plutonium could be used beneficially as fuel. Uranium also is recovered in reprocessing, and in the case of spent fuel from LWR's it contains a somewhat higher fraction of U-235 than natural uranium (between 0.8 and 1.0 percent, compared to 0.71 percent). The effect of using both the plutonium and the recovered uranium in new fuel is to extend uranium resources by about 20 to 30 percent. Whether this will be economically worth-

*The spent fuel from the gas-cooled reactors used in the United Kingdom cannot be stored for any considerable period, and thus requires chemical reprocessing, including plutonium separation, but not recycling.

while will depend on the costs of reprocessing, and also on the costs of uranium and of enrichment. It was widely assumed a few years ago that recycling would be economically attractive; however, estimates of the cost of reprocessing have increased to the point where it now seems possible—and perhaps likely—that this will not be the case.

Nevertheless, for at least two reasons commercial plutonium separation seems almost certain to occur eventually (although it might be restricted to no more than a few highly industrialized countries). The primary reason is that plutonium recycling is essential to the "breeder economy" that seems likely to develop fairly rapidly, at least in Europe if not in the United States.* Second, by separating from the spent fuel nearly all of the plutonium contained therein, and "burning" it as fuel in reactors, the problems of long-term waste management might be simplified considerably. Plutonium could be the most troublesome ingredient remaining in spent fuel after several centuries. Some countries, such as the Federal Republic of Germany, already require plutonium separation prior to waste disposal.

Although there may be less of an immediate incentive for nations to acquire fuel-reprocessing facilities than enrichment facilities, in some respects the problems of trying to induce forebearance will be similar. The costs of reprocessing will be small compared to the total cost of power generation (that is, on the order of 1 or 2 percent). It follows that a nation driven to acquire such a facility by either a commitment to breeders or concerns about waste management is unlikely to be much influenced by offers of subsidies or by lower costs that might be achieved through the use of larger reprocessing facilities available elsewhere. The economy of scale is not likely to be very significant, and may in the past have been exaggerated. Other things being equal, large plants may be able to perform reprocessing at costs lower than small ones, but this advantage may well be offset by different national standards for environmental protection. It is quite possible that nations with moderate power programs could perform their own reprocessing in small plants at costs that could not be matched by large plants in the United States or elsewhere if the large plants were subject to more stringent safety and environmental standards.

But in important ways the issue of reprocessing is different from and more troublesome than that of uranium enrichment, for there is reason for concern about what is done with the plutonium extracted from spent fuel, regardless of where the reprocessing has occurred. Assuming that reprocessing is performed at a multinational or other "approved" facility, the cleanly separated plutonium might be returned to the country of origin, there to be fabricated

*There exists the possibility of an alternative breeder cycle in which uranium-233, a fuel material, is produced from thorium. It is likely that breeders based on this cycle, if they are built at all, will follow those based on the uranium-238, plutonium cycle. In any case, from a proliferation perspective the problems for both systems would be similar since the uranium-233, which would be separated from unconsumed thorium and fission products, could, like plutonium, be used for weapons.

into fuel for use in that country's reactors. But there are a number of objections to this course. The possibility of the material being used for weapons manufacture would in this case be no less than if the reprocessing occurred in the country of origin. Indeed, there would be an added hazard because of the greater possibility of theft or accident in transit. One advantage from the nonproliferation perspective of having the reprocessing done outside worrisome countries is that there would be no danger of "nationalization" of the facilities and any stockpiles of plutonium within them. In the case of a country with a modest nuclear power program, the amount of plutonium available at any time for weapons might in this way be limited to several hundred kilograms (that is, to the amount that might be used in one or two mixed oxide fuel reloads). But this would be enough material to produce tens of weapons. Thus, the provision of plutonium to worrisome countries—even if separated from their own spent fuel—can hardly be categorized as a nonproliferatory measure.

There are two other approaches to dealing with plutonium that merit further discussion. One involves inducing nations to forego the return of their plutonium, and the other involves the return of the plutonium in a form that would make it difficult to use for weapons.

An approach in which the plutonium is not returned to the country of origin could have substantial appeal to countries not operating, or intending to operate, breeders. The plutonium could be sold for use by others, or they could get uranium of equal value (in most cases enriched) in exchange. The latter possibility suggests that there may be an advantage to combining reprocessing, enrichment, and possibly uranium mining and milling under the same multinational management. If guarantees of stability of exchange of uranium for plutonium were sufficient, some nations that might otherwise choose to build breeders could probably be persuaded not to or at least to delay doing so. This approach might well be acceptable to quite a number of states, but very likely would be unacceptable to those highly industrialized nations that do not have uranium reserves and are already interested in breeders, notably Germany and Japan.

The possibility of returning plutonium to states of origin in a form not easily usable for weapons manufacture seems less promising. A number of proposals for accomplishing the task have been made. Prominent among them are: (1) incomplete separation during reprocessing, leaving the plutonium intimately mixed with the much larger amounts of depleted uranium; (2) return of the plutonium in the form of mixed oxide fuel elements in which plutonium and uranium oxides are intimately mixed; and (3) "spiking" the plutonium by adding to it (or not removing during reprocessing) one or more radioactive isotopes which emit high-energy gamma rays. The effectiveness of each of these approaches would be enhanced if the material were returned only as needed for use as reactor fuel, thereby minimizing the quantity available at any time for weapons.

The first approach—dilution—would reduce the threat of weapons being made from plutonium diverted or stolen during shipment because dilution with uranium would render the plutonium unsuitable for use directly in weapons and would mean much larger amounts of material would have to be handled. Such dilution, however, would not represent a substantial barrier to weapons manufacture by the state to which the plutonium was being returned. Any nation having the capability to transform the plutonium-uranium mixture into fuel elements would also have (or could rapidly acquire) the capability to separate the plutonium from the uranium.

The second approach—the return of mixed oxide fuel—is compatible with expected industrial practice. But, again, separation of plutonium from uranium would not represent a significant technical obstacle to the recipient nation.

The third approach—"spiking"—appears attractive, at least theoretically. It would make weapons manufacture somewhat more difficult because of the need for remote handling equipment either to remove the gamma-emitting materials prior to weapons fabrication or to fabricate the weapons using highly radioactive materials. It would also make diversion or theft during transport more difficult and more easily detected. There are, however, serious drawbacks as well. If fuel fabrication were carried out in the country of origin and if the normal operating procedure were to remove the gamma emitters prior to fuel fabrication, then there would be no greater impediment to weapons manufacture by that country than if the separation process itself had been accomplished there. On the other hand, if the gamma emitters were not removed, fuel fabrication and handling would be more difficult and expensive.* To induce recipient nations to adopt the more costly practices would, presumably, require subsidization. In any event, spiking would increase the health hazards of the nuclear industry.

It was suggested previously that some nations might be induced to forego indigenous enrichment facilities by linking enrichment elsewhere to guarantees of uranium supply. Similar arrangements could be applied to the back end of the fuel cycle. Just as uranium is very unevenly distributed, so too are the geological formations in which waste products may be safely deposited. Moreover, there may be other objections on the part of some countries to having wastes buried within their borders. Such countries might find reprocessing abroad very attractive if, as part of the arrangement, they could have their radioactive waste taken off their hands. The question is whether others, such as the United States, the Soviet Union, and Canada, which are relatively well endowed with possible disposal sites, would accept wastes other than their own. At issue would be how effective such linkage could be in limiting the spread of reprocessing technology and how much of a price they are willing to pay to achieve it.

*There is the problem that no gamma emitter of suitable characteristics—including availability at reasonable cost—has yet been identified.

Although this paper has focused primarily on the technical and economic factors relating the nuclear fuel cycle to the proliferation of nuclear weapons, it is clear that the proliferation problem is fundamentally a political one. Searches for a solution in the form of a "technical fix" or an "irresistible economic inducement" are bound to end in failure. If a solution exists, it undoubtedly is a concoction of technical, economic, and political ingredients. Among the tests that must be applied to any proposed solution are:

1. Would it retard (or at least not accelerate) the spread of technologies for the production of nuclear weapons?
2. Would it lengthen (or at least not shorten) the time required from the decision to acquire a weapons capability to the achievement of that capability?
3. Would it raise (or at least not lower) the threshold of the decision to acquire a nuclear weapons capability?
4. Would it decrease (or at least not increase) the dangers of theft and sabotage?
5. Would it support (or at least not undermine) other efforts to halt the spread of nuclear weapons, such as increasing incentives to subscribe to the NPT, strengthening the safeguards effort of the IAEA, negotiating among the supplier nations agreed restrictions on exports, establishing nuclear free zones, and pursuing a comprehensive test ban?

These tests are difficult ones, and few if any proposed solutions will pass them all. But the reward for success is sufficiently great to command a continued search.

Chapter 2

Why Reprocess?

Ted Greenwood

Until recently reprocessing—the extraction of uranium and plutonium from irradiated fuel—and the recycling of these materials through reactor fuel cycles was expected to be profitable. That is, the value of the materials recovered was expected to be higher than the cost of recycling. The light-water reactor (LWR) fuel cycle was designed with that expectation in mind and with the firm belief that sooner or later reprocessing and recycling would take place. Today that expectation has disappeared. Virtually no one today believes reprocessing can be profitable in this sense. The prospective costs of reprocessing, waste management, and mixed oxide fuel fabrication appear to have risen too much. The back end of the fuel cycle is now recognized to be a net cost and not a net credit to the ultimate price of electricity. Until such time as the price of enriched uranium rises high enough to reverse this, the real economic question for the back end of the fuel cycle is whether the cost of reprocessing and recycling is higher or lower than the cost of refraining from doing so. The alternative is permanently to dispose of spent fuel or, because of the value of its fissile content, more likely to store it until the economics of recycling change.

For heavy-water-moderated, natural uranium reactors the current economics are very unfavorable to reprocessing. This is because the fuel can be stored without great difficulty, the amount of plutonium per kilogram of fuel is low compared to LWR fuel, and the uranium that would be recovered is depleted in fissile uranium-235 relative to natural uranium. Canada, the major user of

This paper is adapted from parts of Ted Greenwood, George W. Rathjens and Jack P. Ruina, "Nuclear Power Technology and Nuclear Weapons Proliferation" *Adelphi* 131. London. International Institute for Strategic Studies, 1976. The author gratefully acknowledges the permission of George Rathjens, Jack Ruina and the International Institute for Strategic Studies for permitting the material to be used in this form.

heavy-water reactors, has postponed its own entry into the reprocessing business until the domestic supply of uranium falls or the price of uranium rises sufficiently to make recovery attractive.

For LWR's the calculation is extremely uncertain today. The difference in cost between recycling uranium and plutonium on the one hand and storage of spent fuel on the other is likely to be small given current projections for the price of uranium enrichment, reprocessing, recycling, waste management, and spent-fuel storage. In either case the back end of the fuel cycle is unlikely to account for more than 10 percent of the total fuel-cycle costs and a few percent of total electricity costs. So the decision whether or not to include reprocessing in the light-water fuel cycle can, and probably will, be made by states primarily on grounds other than short-term economics.*

This paper first reviews the current state of the reprocessing industry. It then examines the incentives and disincentives other than short-term economics that will affect the decision of states to acquire their own reprocessing facilities. Finally it examines the possibility of avoiding a widespread commercial reprocessing industry.

THE INTERNATIONAL REPROCESSING INDUSTRY

Table 2-1 provides a list of all nonmilitary reprocessing facilities available, under construction, or planned. As can be seen from the table, many nations have already started down the path toward commercial reprocessing.

The United States

The first reprocessing facility was built during World War II at Hanford, Washington, to produce plutonium for the Manhattan Project. This initial facility, using a chemical precipitation process, was replaced after the war by solvent extraction plants at both Hanford and Savannah River, Georgia. These plutonium-production reprocessing facilities are designed to handle fuel elements fabricated of natural uranium metal, clad with aluminum (Hagerton, 1963: 191-93, 219-20, 492-93). The Idaho Chemical Processing Plant at the National Reactor Testing Station is used as a production facility, an engineering-scale process-demonstration facility, and an operational facility to reprocess naval reactor cores (USAEC, 1971: 17-19). The Metal Recovery Plant at Oak Ridge National Laboratory was the initial pilot plant for reprocessing thorium fuels.

The American commercial LWR reprocessing industry was initiated in 1966 by the Nuclear Fuel Services facility at West Valley, New York. In its nearly six years of operation it recovered, not 300 tonnes per year as designed, but only

*The gas-cooled reactors, used only in the United Kingdom and France, are a special case. The unstable nature of the irradiated fuel makes immediate reprocessing preferable, and this has been under way in those countries for many years.

Table 2-1. Nonmilitary Reprocessing Facilities

Country	NPT Party	Facility	Fuel Type	Design Capacity	Safeguards	Status
Argentina	No	Ezeiza Nuclear Facility	Metal (research-reactor fuel)	Lab scale	Yes	Shut down, but being reactivated for operation in 1977; may include redesign for low-enriched UO_2 fuel.
Belgium	Yes	Eurochemic, Mol	Metal and UO_2 low enrichment, and metal high enrichment	60–85 tonnes low enr./yr; 1.25 tonnes high enr./yr	Yes	Start-up 1966; shut down 1974.
Brazil	No		UO_2 low enrichment		Yes	Purchase from Federal Republic of Germany.
France	No	Cogema, La Hague (United Reprocessors)	Natural U metal	800 tonnes/year	No	Start-up 1967.
			Low enrichment UO_2 head end	800 tonnes/year	No	Start-up in May 1976; 800 tonnes/yr by 1982
		Cogema	Low enrichment UO_2	1000 tonnes/year	No	Under consideration.
		Cogema, Marcoule	Natural U metal	1000 tonnes/year	No	Start-up 1958 for military purposes; will take over commercial role from La Hague.
			LMFBR	5 tonnes/year pilot plant	No	Operational.
West Germany (Federal Republic)	Yes	KEWA (United Reprocessors)	UO_2 low enrichment	1500 tonnes/year	Yes	Planning stage, start-up 1986 or later.
		KEWA, Karisruhe	Breeder and UO_2	200 kg/day pilot plant	Yes	In operation since 1971.
		KFA, Juelich	Graphite	2 kg/day pilot plant	Yes	Start-up 1977.

(continued)

Table 2–1 continued

Country	NPT Party	Facility	Fuel Type	Design Capacity	Safe-guards	Status
India	No	Trombay	Thorium/Uranium Oxide	Lab scale	No	In operation.
		Trombay	Metal and UO_2	350 kg/day	No	In operation.
		Tarapur	Metal and UO_2	100 tonnes/year	No[a]	Being cold tested.
Italy	Yes	Eurex-1, Sallugia	UO_2 and metal	Small pilot plant	Yes	In operation.
		ITREC, Rotondella	Thorium/Uranium	Small pilot plant	Yes	In operation.
		Unnamed	UO_2 low enrichment	500 tonnes/year		Proposed start-up 1985; plans temporarily shelved.
Japan	No	PNC, Tokai-Mura	UO_2 low enrichment	200 tonnes/year	Yes	Tested, closed for alterations; commercial operations late 1977 or beyond.
		PNC, Tokai-Mura	Metal	Small plant	Yes	In operation.
Pakistan	No	unknown	UO_2 low enrichment	unknown	Yes	Purchase from France.
Spain	No	Juan Vigon Ctr., Madrid	Metal	Small pilot plant	Yes	In operation.
Sweden	Yes		UO_2 low enrichment	800 tonnes/year	Yes	Under consideration.
Taiwan	No	Nuclear Energy Research Institute	Metal	Lab scale	Yes	Being built.
U.S.S.R.	Yes				No	In operation.
United Kingdom	Yes	Dounreay	MTR and fast reactor fuel	Small pilot plant	No	In operation.
		British Nuclear Fuels Ltd., Windscale (United	Natural Uranium Metal	2000 tonnes/year	No	In operation.

		Reprocessors)	Low enrichment UO_2 head end	Small scale commercial	No	Operated 1969–73; shut down for modification; will reopen late 1970s to feed into natural uranium separation plant depending on availability of capacity.
			UO_2 low enrichment	1000 tonnes/year	No	Start-up mid-to late 1980s for expected domestic and overseas requirements.
United States	Yes	Nuclear Fuel Services, West Valley	UO_2 low enrichment	300 tonnes/year	No	Operated 1966–1972; closed for expansion and rebuilding.
		W. Valley	UO_2 low enrichment	750 tonnes/year	No	Early 1980s.
		General Electric, Morris	UO_2 low enrichment	300 tonnes/year	No	Inoperable in present form.
		Allied General Nuclear Services, Barnwell	UO_2 low enrichment	1500 tonnes/year	No	In construction; operation expected 1978 or after.
		Exxon Nuclear, Oak Ridge	UO_2 low enrichment	1500 tonnes/year	No	Application pending.
		ERDA and General Atomic, Idaho Falls	HTGR fuel	Demonstration scale	No	Under consideration.
		ERDA	LMFBR fuel	Demonstration scale	No	Under consideration.
Yugoslavia	Yes	Boris Kidric Metal Institute		Lab scale	Yes	In operation.

[a]Will be safeguarded when reprocessing American-supplied fuel.

630 tonnes (Clark, 1976: 27-31). In 1972 it closed for major rebuilding and enlargement and, because of regulatory delay, is not expected to reopen before the early 1980s. Efforts to begin operations at General Electric's 300-tonnes-per-year pilot plant at Morris, Illinois, based partially on new technology, were abandoned in 1974, after three years of trying and an investment of $64 million, because of technical problems and design inadequacies (Gillette, 1974: 770-71). Cost of reconstruction has been projected at more than $130 million and to require four or more years following a decision to begin ("G.E. Decision . . . , 1975: 59).

The 1500-tonnes-per-year Allied General Nuclear Services plant at Barnwell, South Carolina, has also been delayed because of regulatory uncertainty. These delays and the addition of ancillary facilities to convert plutonium nitrate solution to plutonium oxide and to solidify high-level wastes have driven construction cost estimates from the original $70 million to $600 million and above (Rogers et al., 1975: 2; Tannenbaum, 1976: 1). Financial assistance from the federal government is being sought to help the company deal with these escalating costs. Start-up is not expected before 1978 and may well be delayed further.

The only other commercial LWR reprocessing facility under active consideration in the United States is a 1500-tonnes-per-year plant of Exxon Nuclear at Oak Ridge, Tennessee. An application was sent to the Nuclear Regulatory Commission in early 1976 (*Nucleonics Week,* Feb. 5, 1976: 2; *Nuclear News,* Mar. 1972: 59). Planning had been in progress for the construction of a demonstration reprocessing facility for fuel from high-temperature gas-cooled reactors and large enough to service ten to twenty reactors. It was once expected to be built in the 1980s by government and industry and to be operating by the early 1990s. However, the great uncertainty about the future of the high-temperature gas-cooled reactor industry in the United States has put this plan in doubt (ERDA, 1975). A demonstration reprocessing plant for liquid metal fast breeder reactor fuel is anticipated for the 1980s or 1990s (ERDA, 1975: 153).

Much of the delay and cost escalation for the U.S. reprocessing industry is directly attributable to regulatory changes and delays. The rules concerning construction specifications, plant siting and safety, transportation of nuclear fuel, waste storage, and physical protection have all changed while the Barnwell plant has been under construction. The major uncertainty still remaining is the acceptability of using mixed oxide fuel in LWR's. A decision from the Nuclear Regulatory Commission is now expected in 1977 but may be delayed beyond that (*Nuclear News,* Dec. 1975: 17). Assuming a positive decision and in the absence of additional delays because of court challenges or congressional action (which are, in fact, quite likely), the industry might be close to normal commercial operations by the early 1980s. But no facilities beyond the Barnwell, West Valley, and perhaps the Morris plants will be operational before the mid-1980s. New investment from private industry will probably await not only a clarification of the regulatory environment, but also a demonstration of the profitability of existing ventures. In the meantime, both the accumulation and the annual output of spent fuel will continue to rise.

Industrialized Countries

The prospects of the commercial reprocessing industry are better outside the United States (Rippon, 1976: 21-27; Rossney, 1976: 20-21; U.S. Senate, 1975: 27). The French have operated two natural uranium metal plants for many years and are now converting one to accept low, enriched uranium oxide fuel. Operations began in May 1976 and full capacity of 800 tonnes per year is expected by 1982 (*Nuclear Engineering International,* July 1976: 8). Design work has begun on a larger uranium oxide facility expected to be completed in the mid-1980s. A small pilot plant for liquid metal fast breeder reactor fuel is also operational.

Uranium metal reprocessing of gas-cooled reactor fuel has been carried out at Windscale in the United Kingdom for twenty-four years. In 1968 a 400-tonnes-per-year head-end plant was added to accept oxide fuel. It operated until 1973 when it was closed down because of technical difficulties. This is being refurbished and is expected to reopen in the late 1970s. A new 1000-tonnes-per-year plant is under consideration for the mid- to late 1980s. It would service the domestic market and, pending a decision that importing foreign fuel for reprocessing is socially acceptable, it would serve the overseas market as well. Britain has a long tradition in international sales of reprocessing services. It has accepted fuel from the gas-cooled reactors in Italy and Japan and smaller amounts of fuel from Switzerland, West Germany, Spain, Sweden, Canada, and Belgium (*Nuclear Engineering International,* Dec. 1975: 989).

Under the plans of the Anglo-French-German joint company, United Reprocessors, which is coordinating the growth of uranium oxide reprocessing capacity in the three countries, a 1500-tonnes-per-year facility is being planned for operation in West Germany by 1986 (*Nuclear News,* June 1976: 63). Significant political difficulties are anticipated in selecting a site, however. In the meantime, Kernbrestoff-Wiederaufarbeitungs-Gesellschaft mbH (KEWA), the German partner in United Reprocessors, is operating a 40-tonnes-per-year uranium oxide pilot plant at Karlsruhe. A small graphite fuel pilot plant is also under construction in Germany and is expected to begin operation in 1977.

The Eurochemic pilot plant at Mol, Belgium, operated for eight years but closed in 1974. A Belgian group is considering expansion of this plant to 300 tonnes per year and reopening it in the late 1970s. The Italians had planned to build a large commercial facility by the mid-1980s, but these plans are now reported to be shelved. Two small pilot plants are operational in Italy, one for thorium/uranium fuel and the other for uranium oxide or uranium metal. Sweden has under consideration an 800-tonnes-per-year facility anticipated for the early 1980s (*Nuclear Engineering International,* July 1976: 23).

A 200-tonnes-per-year uranium oxide Japanese facility has been tested and shut down for alterations. Commercial operation will be delayed beyond early 1977 (*Nuclear Engineering International,* June 1976: 12). Plans to build a commercial-scale plant adjacent to the pilot plant are running into considerable opposition. In Japan, as in Germany, a utility must demonstrate contractual commitments for reprocessing services before a construction permit for a nuclear power plant will be issued. The Japanese have been seeking contracts in the United Kingdom and France for the 1980s (*Nuclear News,* May 1975: 63).

Developing Countries

Two metal and low enrichment uranium oxide facilities have been built in India. One is in operation and the other is now being tested with nonradioactive fuel. The Indians also have a small laboratory-scale thorium/uranium facility. Taiwan is building a lab-scale metal fuel facility. Spain and Yugoslavia have small pilot plants in operation. Argentina's facility has been shut down for expansion and redesign and is expected to reopen in 1977. Brazil has contracted with Germany for the purchase of a small plant. The Republic of Korea has recently been dissuaded by the United States from pursuing its negotiations with France for the purchase of a pilot facility (Halloran, 1976: 11; *Nucleonics Week,* Feb. 5, 1976: 4). However, France's sale of a plant to Pakistan seems likely to be accomplished despite intense U.S. pressure against it.

INCENTIVES AND DISINCENTIVES FOR NATIONAL REPROCESSING

Incentives

Although the recycle of uranium and plutonium may not be economically attractive now, that situation may not last forever. The size of the world's ultimate uranium reserves at a cost that makes retrieval economically feasible (say $100 per pound) and the future rate of growth of uranium demand are both extremely uncertain. Nonetheless, most projections of demand and supply indicate that currently known reserves would be exhausted before the turn of the century and that currently estimated additional reserves, if they turn out to be real, might be committed under long-term contract before then (OECD, 1975). As the price rises and supply becomes tighter, the economic attractiveness of reprocessing and recycling will improve. In the meantime it would reduce the dependence of uranium-short states on foreign supplies, limit their vulnerability to supply curtailment, and ease the pressure on their balance of payments. This effect is not very strong, however, since recycling uranium and plutonium in LWR's reduces the demand for natural uranium and enrichment services by only 20 to 30 percent.

For some states, dealing with spent fuel is not primarily a matter of economics. They are simply reluctant to keep it around and feel that if reprocessing will become necessary or attractive eventually, it might just as well be done immediately and the final waste products put into permanent geologic storage. Such a country will want access to a reprocessing plant primarily as a means of nuclear waste management.

Another reason for some nations to be interested in reprocessing is their anticipation of a breeder economy some time in the future. For nations with a large commitment to nuclear power and limited uranium resources, breeder reactors are likely to be seen as necessary and inevitable. France, the United Kingdom, and the Soviet Union already have small breeder reactors operating. Japan is not far behind. It is possible that their first commercial units will be operational within a decade. Germany and the United States also have programs under way. The rate of diffusion of the technology both within the nations that are developing it and to others is very uncertain. Some countries may want to purchase breeders as they have purchased the current generation of reactors and may not want to wait long after feasibility has been proven. At least they may want to take actions in the meantime that provide that option. To run a breeder fuel cycle, reprocessing is essential. Moreover, to initiate the operation of a breeder reactor, a large supply of plutonium or highly enriched uranium will be necessary. Some states will want to reprocess their spent fuel from current reactors and recycle some plutonium in order both to gain experience in working with mixed oxide fuels and to accumulate a plutonium stockpile on which to draw when breeders are eventually introduced.

These inducements for states to want access to reprocessing facilities are more weighty in the long run than in the short run and could be accommodated through access to reprocessing facilities in other countries. Nevertheless, there are a number of other reasons that might incline a nation to acquire its own facilities and to do so soon. Any country that invests the large sums of money required to initiate and nurture a nuclear power industry does so primarily because it thinks nuclear power can provide a secure and economic source of usable energy over the long term. It will inevitably be concerned that fuel-cycle services required to operate its reactors reliably, efficiently, and with minimum environmental cost will be available. Reliance on others can be reasonably dependable if long-term contracts are available, if supplier states have political or unrelated economic reasons not to default, and particularly if some overcapacity exists and several suppliers are vying for business. Nonetheless, some states may want their own facility strictly out of concern for security of supply of reprocessing services. Even a small facility will serve the purpose, at least until the domestic nuclear sector grows. While perhaps more expensive than purchasing services from a large, more efficient plant, states are often willing to pay economic penalties to achieve greater security for important commodities and services.

Government officials will no doubt also be impressed by the extended time required to build and learn to operate even a small, pilot reprocessing plant. The lead time until a large-scale facility could be fully operational is considerably longer. With the uncertainties about uranium supply and the availability of enrichment services, the small relative cost of reprocessing compared to the cost of commercial reactors themselves and the apparent speed with which at least a few countries are moving ahead with breeder technology, some nations will conclude they should move ahead now rather than wait.

A few countries may see economic benefit or commercial opportunity in a domestic reprocessing industry. Not only might they save foreign exchange by not purchasing services abroad, but they might also earn foreign exchange and generate employment opportunities by selling services to others. Britain has done this and seems interested in expanding its role as an international vendor of reprocessing services. Still other states will insist on acquiring their own plants simply because their vision of their own national *grandeur* or ambitions requires that they both possess and demonstrate the ability to build strategically important industries of which reprocessing is a prime example.

Finally, some states will see benefit in acquiring a means of access to plutonium and thereby reducing the lead time required to build nuclear explosives if they should decide to do so. This benefit may be related to an active interest in nuclear weapons, to a desire to open options for the future, or to a desire to seek political benefit by engendering ambiguity in the perceptions of others.

Disincentives

The list of possible incentives for states to reprocess spent fuel and even to acquire a domestic reprocessing plant is lengthy. There are also disincentives, however. First there are the health and safety risks intrinsic to working with or transporting such a toxic material as plutonium. Second is the possible reluctance to risk the theft of plutonium and subsequent use or threatened use of nuclear weapons by terrorists or other non-state entities. Third is the cost of building a reprocessing plant. In countries with scarce capital or skilled personnel, the fact that a reprocessing plant represents only a small percentage of a full nuclear industry may not be sufficiently persuasive to overcome resistance to investment that is less than essential for economic development. Fourth, there may be political costs to building a facility that is closely related to weapons options. Acceptance of International Atomic Energy Agency safeguards and even becoming part to the Non-Proliferation Treaty may not sufficiently ameliorate the adverse reaction of neighbors, adversaries, or allies.

CONCLUSIONS

The manner in which all these incentives and disincentives will work out to an ultimate choice will clearly vary from state to state, and for any particular state may vary over time. To many observers the primary objective of nonproliferation policy is to encourage as many states as possible to forego reprocessing in the short run and to extend as long as possible the length of time they are willing to do so. The foregoing analysis suggests several necessary components of such an effort. Success in the short run will require both the discovery of sufficient uranium supplies to eliminate or reduce the apparent need to recycle fissile material from irradiated fuel and to develop plutonium breeders, and the demonstration of acceptable and economic means of storing spent fuel. For the longer run either termination of plutonium breeder development programs or, at the least, an announced policy of restraint in marketing them internationally, will also be required. Assuming the fulfillment of these requirements many states with nascent and perhaps even substantial reactor programs might decide not to reprocess their spent fuel or at least to forego doing so for some time. In the more likely event that these requirements remain unfulfilled, the number of such states will be small.

In either case some states might prefer, be willing, or be persuaded to rely on reprocessing services provided by others. To encourage this latter route two approaches could be taken by the industrialized countries or the international community. The first would be the creation of sufficient reprocessing capacity in nuclear or low-risk non-nuclear states to assure competition among suppliers and therefore to increase the security of access by others to reprocessing services at attractive prices. The second—not necessarily mutually exclusive—alternative would be the creation of the multinational fuel-cycle facilities that are the subject of this volume. From the point of view of addressing the incentives identified here the second approach would seem advantageous only to the extent that participation in the ownership of a multinational facility would satisfy symbolic and prestige objectives that otherwise might encourage states to build their own national facilities.

Both approaches suffer from one serious inadequacy as a nonproliferation mechanism. If purified plutonium (or plutonium compounds) are returned immediately to the country of origin either for recycling in current reactors or for stockpiling to await breeder reactors, little would have been gained over the alternative of national reprocessing facilities. Some provision for the reprocessor to keep the plutonium, whether simply as a storage agent for the owning state or in exchange for payment in money or enriched uranium, would therefore enhance the value of either multinational facilities or a buyers' market composed of essentially national plants. Nations wishing eventually to move into breeders would want to retain ownership of all or some of their plutonium, and those wishing to gain experience in the use of mixed oxide fuels would want to reclaim at least some portion before breeders are available. To accommodate these requirements plutonium could be returned by the reprocessor as needed, under appropriate safeguards. In this way at least national plutonium stockpiles would be precluded.

Those countries that are primarily concerned about waste management may have no desire to recover their plutonium, at least for some time. As already suggested, any scheme that removes spent fuel for storage or ultimate disposal of wastes would provide for them a major incentive to forego national reprocessing. An institutional arrangement to supply this service could be national, multinational, or international, and might include reprocessing, but could be—at least in the short run—only a spent-fuel repository.

Even in the presence of spent-fuel repositories, multinational facilities, and a buyers' market, there will no doubt be nations that will still insist on building their own facilities. Their own motivation could be a residual anxiety about security of supply, a reluctance to spend scarce foreign exchange to purchase services elsewhere, a concern about national prestige, or the desire to create a weapons option. The competitive nature of the international nuclear industry and the potential entry of a growing number of states into the marketing of reprocessing technologies make the success of efforts to impose an effective export embargo on reprocessing technology unlikely. Even if an embargo were imposed, however, the fact that reprocessing technology is not shrouded in secrecy and the nature of the technology itself mean that any state able to sustain a nuclear or chemical industry and willing to invest several tens of millions of dollars could build a small commercial plant on its own. A facility not tied to a commercial nuclear industry, but whose only purpose was the recovery of plutonium from low-burnup fuel for an explosives program, would be significantly less difficult to build. Short of political or military coercion, therefore, there is little that can be done to prevent a state from acquiring its own reprocessing plant if dissuasion through market manipulation fails.*

REFERENCES

Clark, J.R. "Modifying the West Valley Reprocessing Plant," *Nuclear Engineering International,* (February 1976).

Energy Research and Development Administration (ERDA). *A National Plan for Energy Research, Development and Demonstration: Creating Energy Choices for the Future.* ERDA-48, Vol. 2. Washington: Government Printing Office, 1975.

"G.E. Decision of MFRP to Await NRC Policy," *Nuclear News* (September 1975).

Gillette, Robert. "Nuclear Fuel Reprocessing: GE's Balky Plant Poses Shortage," *Science,* 30 August 1974.

*The implications of this ease of building reprocessing plants and the apparent inevitability of the diffusion of the reprocessing technology and facilities, for overall nonproliferation strategy is beyond the scope of this paper. For the author's views on this broader subject, see "The Proliferation of Nuclear Weapons," a paper presented at the 18th Annual Conference of the International Institute for Strategic Studies; or "Toward a Strategy for Non-Proliferation," to be published by The 1980s Project of the Council on Foreign Relations.

Hagerton, John F. *The Atomic Energy Deskbook.* New York: Reinhold, 1963.

Halloran, Richard. "Seoul Officials Say Strong U.S. Pressure Forced Cancellation of Plans to Purchase a French Nuclear Plant," *The New York Times* February 1, 1976, p. 11.

Nuclear Engineering International, various issues as cited.

Nuclear News, various issues as cited.

Nucleonics Week, various issues as cited.

OECD, Nuclear Energy Agency and the International Atomic Energy Agency. *Uranium Resources, Production and Demand.* Paris: OECD, 1975.

Rippon, S., Reprocessing–What Went Wrong? *Nuclear Engineering International,* Feb. 1976.

Rogers, Walton A.; Pickard, James K.; and Vallence, John W. "The Throwaway Fuel Cycle," Paper presented to the Atomic Industrial Fuel Cycle Conference, 75 Atlanta, March 20, 1975.

Rosmay, G., Reprocessing in Europe, *Nuclear Engineering International,* April/May, 1976.

Tannenbaum, Jeffrey A., "White Elephant? Big Plant to Recycle Nuclear Fuel Is Hit by Delays, Cost Rises," *The Wall Street Journal,* February 17, 1976, p. 1.

United States Atomic Energy Commission, *National Nuclear Reactor Testing Station,* July, 1971.

United States Senate, Committee on Government Operations, *Facts and Nuclear Proliferation,* 94th Cong., 1st Sess. (December 1975).

Part II

Technical Aspects

Chapter 3

Technical and Economic Considerations in Fuel Reprocessing

S.R. Hatcher and W.W. Morgan

For the remainder of this century the bulk of the world's nuclear generating capacity will be in uranium-burning thermal reactors. Figure 3-1 shows the estimates for world nuclear capacity derived from studies made for OECD (Crawford & Husserman, 1975) and Japan (Takei, 1975) in November 1975. By the year 2000, 84 percent of the world's capacity is expected to be in enriched uranium reactors, 8 percent in natural uranium reactors, and 8 percent in fast reactors.

By the mid-1990s, it is assumed, fast reactors will start to penetrate the market in the U.S. and Western Europe. Natural uranium reactors are expected to predominate in Canada and some developing countries, while enriched uranium reactors (principally LWR's) will be used in all other countries.

Figure 3-2 shows the spent fuel available annually for reprocessing from thermal reactors. By the year 2000, the natural uranium fuel accounts for 26 percent of the total because of its low burnup (7500 MWd/MgU vs. 30,000 MWd/MgU for enriched reactors). With their much higher fuel burnup, the fast reactors will account for less than 4 percent of the fuel available for reprocessing at the end of the century. Thus there may be fuel available for 25 to 30 LWR fuel and 10 natural uranium reprocessing plants.

Within this time scale then, we are concerned principally with large-scale fuel-cycle centers for handling enriched uranium or natural uranium from thermal reactors. The reprocessing operations involve the separation and recovery of plutonium and uranium from fission products.

REPROCESSING TECHNOLOGY

Two types of reprocessing methods are available—aqueous and nonaqueous. The aqueous processes are the only ones which have advanced to industrial practice,

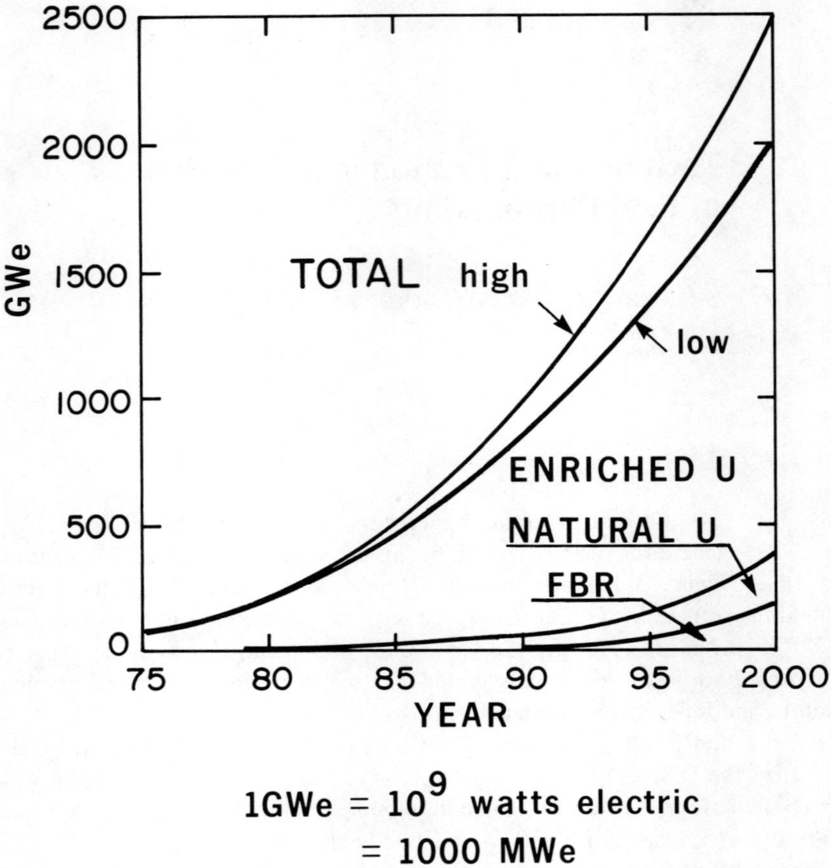

Figure 3-1. World Nuclear Capacity (November 1975 estimates).

and the major one—solvent extraction—is discussed in some detail below. Of the nonaqueous processes, fluoride volatility was developed to the highest degree. The process involves fluorination of uranium and plutonium oxides to the gaseous UF_6 and PuF_6 derivatives. Each component can then be separated and purified by distillation. Pilot tests showed that pure fractions could be prepared. The major attractions to fluoride volatility were: (1) criticality problems were alleviated because no water is present; and (2) wastes do not proliferate because no solid chemicals are added. Unfortunately, fluorine gas handling is hazardous and PuF_6 leakage could present a major emission problem. So further development of fluoride volatility as a primary separation process was suspended in favor of solvent extraction (Jonke, Levitz, & Steindler, 1969).

Solvent extraction is the only process which has been developed to a commercial scale and the only one being seriously developed for new applications.

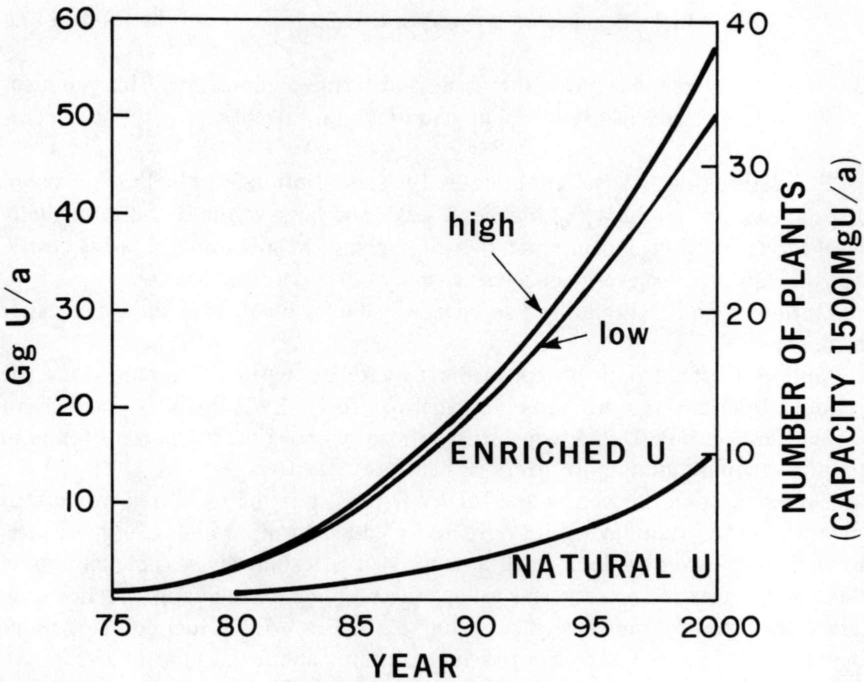

1 Gg U/a = 10^9 grams uranium per annum
= thousands of metric tons per annum
1 Mg U = 10^6 grams uranium
= 1 metric ton

Figure 3–2. World Fuel Available for Reprocessing (Excluding FBR)

The principal advantage of solvent extraction is in the use of liquid streams. This allows a continuous cascade process to be designed and operated easily, and it is the continuous cascade which allows the required separations to be made with high decontamination factors and high product recovery, both of which are essential for an efficient recycle system.

Having selected solvent extraction as the basic process, there are still several options open:

1. Extract only the new fissile components. This could be used for recovery of plutonium from fuels where the U-235 has been depleted below the tails concentration of enrichment plants.
2. Extract the fissile and fertile components together; that is, plutonium with uranium. This option could be applied to LWR fuels but is of little interest

for CANDU fuels, because of the very low U-235 content in the residual uranium.

3. Extract and separate both the fissile and fertile components. This produces two separate streams of plutonium and uranium.

Standard world practice for uranium fuels uses option 3 in the Purex process. It produces small volumes of high-level waste and large volumes of intermediate-level waste. Option 3 plants are the most complex and thus the most costly, but they do produce every conceivable type of product from the fuel.

Options 1 and 2 should lead to cheaper, simpler plants that should be easier to operate.

Option 1 would only be applicable to CANDU natural uranium where the residual uranium has no value for thermal fuels. LWR fuels contain about 0.9 percent residual U-235, which is significantly above the 0.7 percent found in natural uranium; uranium recovery is therefore attractive.

Option 2 could be considered for LWR fuels, but the recovered mixture of uranium and plutonium would have to be added to more highly enriched uranium to produce new fuel. Such an approach is technically feasible and could have advantages in security and safeguards requirements, but in practice such advantages would result in higher fueling costs. Fresh LWR fuel contains about 3 percent U-235, while discharged fuel contains about 0.9 percent U-235 and 0.6 percent fissile plutonium; that is, the fissile plutonium is equivalent to about one-fifth of the fissile content required in new fuel. If the plutonium were separated, it would be used to make up the entire fissile content of a part of the new fuel charge by mixing it with depleted or natural uranium. Such plutonium fuel would then account for about one-quarter to one-fifth of the new fuel charge, the balance being standard enriched uranium. However, if the plutonium were not separated from the uranium, the entire quantity of recovered material would have to be made into fresh fuel by adding further uranium enrichment, at say 5–10 percent U-235. Thus the total amount of plutonium fuel to be fabricated would be four to five times as much as for separated fuel. The cost of producing plutonium fuel is estimated to be 60 percent higher than for UO_2 fuel in a 160 Mg/a plant.

Furthermore, the total separative work requirements in uranium enrichment for the unseparated plutonium fuel case are somewhat higher at about 3.4 SWU/kg new fuel (compare 2.8 for separated Pu fuel and 3.8 for fresh uranium fuel). All these economic penalties would have to be balanced against the apparent advantages in security and safeguards and possible savings in reprocessing costs.

The basic steps in reprocessing are shown in block diagram in Figure 3-3. The first stage is preparation of the feed for extraction. Here the fuel bundles are chopped up to expose the fuel and treated with acid to dissolve the UO_2. Fission product gases are released in the chopper and dissolver and are treated

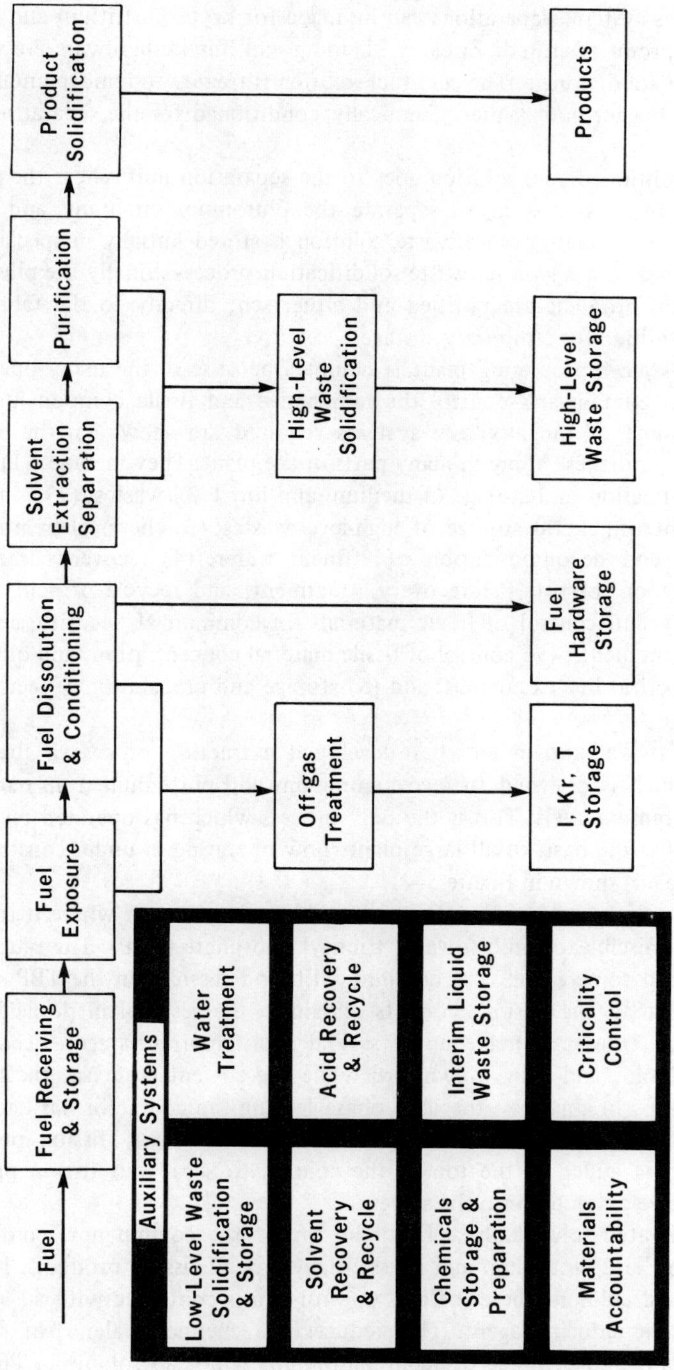

Figure 3–3. Fuel Reprocessing Plant.

in an off-gas system. Separations can be made for krypton, tritium and iodine, and these products stored. Zircaloy cladding and bundle hardware are washed and sent to solid storage. The acid fuel solution is treated to remove undissolved particulate matter and is then chemically conditioned for the separations process.

The conditioned fuel solution goes to the separation unit where the solvent extraction process is used to separate the plutonium, uranium, and fission products. The fission-product waste solution is stored initially in special tanks and then goes to a high-level waste-solidification process. Finally the plutonium and uranium products are purified and either sent directly to the fabrication plant or solidified for temporary storage.

A large-scale reprocessing plant is complex because of the many operations required to contain and control the radioactive and fissile components being handled. Some of the auxiliary systems required are shown on the inset of Figure 3-3, and these apply in many parts of the plant. They include: (1) collection, solidification and storage of medium and low level wastes; (2) concentration and interim liquid storage of high-level wastes; (3) chemical treatment of feed-water and decontamination of effluent water; (4) recovery, treatment, and recycle of solvent; (5) recovery, treatment, and recycle of acid; (6) accountability and control of fissile materials for commercial, security, and safeguards requirements; (7) control of fissile material concentrations and quantities to prevent criticality excursions; and (8) storage and preparation of feed chemicals.

The most well-known and best-developed extraction process is the Purex process, which is designed to recover uranium and plutonium from natural or enriched uranium fuels. This is the only process which has operated on a large scale and it is the basis of all large plants now operating or under construction. The principle is shown in Figure 3-4.

The acid fuel solution is fed to a liquid-liquid contactor where it is mixed with an immiscible organic solvent, tributyl phosphate (TBP). The plutonium, uranium, and some nitric acid combine with and dissolve in the TBP solvent, while almost all the fission products remain in the residual nitric acid. Each contactor or column is made up of several countercurrent stages. Feed enters at the midpoint and flows downward while the solvent enters at the bottom and flows up. In this way the acid phase leaving the contactor has had all its uranium and plutonium extracted and now contains only fission products. Dilute acid is added at the top of the column to scrub out fission products that may have extracted into the solvent.

The separated solvent phase from the top of the column now contains all the uranium and plutonium and is essentially free of fission products. It is fed into a second column where it flows upward and is contacted with an aqueous solution of a reducing agent. This reduces the chemical valence of the plutonium from 4 to 3. Since 3 valent plutonium is not so soluble in TBP, it is

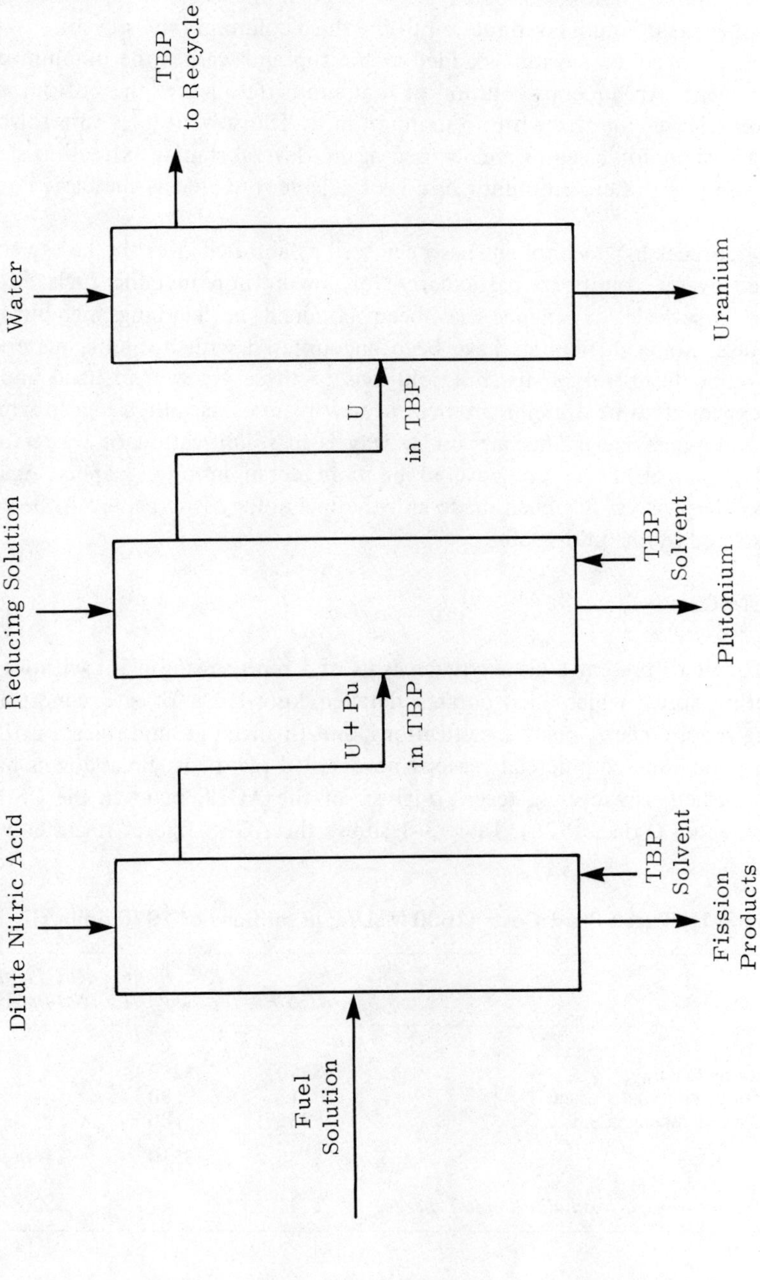

Figure 3–4. Fuel Reprocessing: Purex Primary Separation.

washed out into the aqueous phase where it continues down the column. Fresh TBP is added at the bottom to ensure that all the uranium is kept in the solvent. So the separated plutonium leaves the bottom of the column in an aqueous phase, while the uranium continues on to the third column in the solvent.

In the third column water is added at the top and washes the uranium out of the solvent. An aqueous solution of uranium nitrate leaves the bottom and the solvent leaves the top with no uranium in it. The solvent goes to a solvent recovery system for cleanup and is used again. Several similar extraction steps may be used, and there are minor options available, but that is the basic Purex process.

Fuel reprocessing technology has been well established over the last twenty years in several countries, particularly for low-burnup metallic fuels. More recently large-scale experience has been obtained in handling high-burnup oxide fuels. Some difficulties have been encountered with insoluble materials in the oxide head-end plants, but solutions to these are well in hand and it is not expected to be a major problem in new plants. The other area in which large development programs are under way is in solidification of wastes and emissions control. These are covered in more detail in other papers. Again, considerable progress has been made and the technology is expected to be well demonstrated by the mid-1980s.

ECONOMICS

It is difficult to present a clear cost analysis of a reprocessing plant without an engineering study which incorporates detailed knowledge of site conditions, licensing requirements, plant specifications, and financial ground rules. Furthermore, for obvious commercial reasons no detailed plant-cost breakdowns have been published. However, a recent analysis of the AGNS plant in the US has been presented (Price, 1976). Table 3-1 shows the AGNS figures together with

Table 3-1. Purex Plant Costs (1500 MgU/a; in millions of 1976 dollars)

	AGNS[a]	*RFC Plant (Enriched U)*[b]	*RFC Plant (Natural U)*
Capital Cost			
Separations & UF_6	\$250	\$250	
PuO_2 conversion and storage	115	80	
On-site waste management	231	180	
	\$596	\$510	\$440
Annual Operating Cost (excluding capital charges)	\$ 41	\$ 41	\$ 20

[a]Price, 1976.
[b]AECL estimate.

Table 3-2. Reprocessing Costs for RFC Plants (1500 Mg/1976 dollars)

	Commercial Financing		*Debt Financing*	
Plant life, years	20		20	
Debt/Equity	40/60		100/0	
Debt interest rate	10%		10%	
Return on equity after taxes	15%		–	
Transportation distance	800 km		800 km	
Costs ($/kgU)	*Enriched*	*Natural*	*Enriched*	*Natural*
Capital charges	$108	$93	$48	$41
Shipping	17	6	15	3
Waste handling	26	6	26	6
Operating	27	13	27	13
Totals	$178	$118	$116	$63
Pu fissile (g/kgU)	6.1	2.7	6.1	2.7
U-235, wt	0.9%	0.2%	0.9%	0.2%
Net cost of Pu ($/g)	$7	$44	-$4	$23

an AECL estimate for an "optimized" plant as part of a recycle fuel center in which the plutonium is fed directly to the fabrication plant.

Table 3-2 shows a breakdown of reprocessing costs for plants at recycle fuel centers together with the ground rules assumed for commercial and debt financing. Reprocessing costs range from approximately $60/kgU for natural uranium with debt financing to $180/kgU for enriched uranium with commercial financing.

The following factors should be noted:

1. The capital cost allows for both contact and remote maintenance. In highly active areas, remote maintenance is used while the rest of the plant is assumed to be decontaminable and repaired directly. Experience has shown this philosophy to be sound and it is being generally followed in new plants.
2. Costs for structures to withstand earthquakes and tornadoes are included.
3. Insurance, safeguards, and security costs are included within the present licensing climate which exists in most of the Western world. If national licensing requirements change or safeguards agreements are renegotiated, then costs could be expected to increase or decrease accordingly. The magnitude is difficult to assess but outer limits of ±15 percent on overall reprocessing cost seem appropriate.
4. Tritium and krypton are vented to the atmosphere.
5. Extreme measures to recover the last traces of iodine are not included.
6. Decommissioning costs are not included. However, assuming they are of the same order as decommissioning reactors, then costs are not expected to be greater than $\sim$0.1 mill/kWh, or about +10 percent. This estimate as-

sumes the site will be returned to unrestricted use. If the site remains in the nuclear industry then costs will be less.

The residual uranium value in natural uranium fuel is essentially zero, while the value in enriched uranium (assuming \$63/SWU, tails concentration of 0.2 percent and \$90/kgU as U_3O_8) is \$138/kg. This gives net costs of fissile plutonium of -\$4 to \$7/g for enriched fuel and \$23 to \$44/g from natural uranium. These figures indicate clearly why reprocessing of LWR fuels is still considered an economic advantage, even at current reprocessing costs. Furthermore it should not be concluded that the reprocessing of natural uranium fuels is not economic. Fuel-cycle costs for CANDU reactors are approximately half those of LWR's, and the trend is likely to be maintained. Since heavy-water reactors can use the plutonium more effectively than LWR's (particularly with the thorium cycle), it can be shown that CANDU reactors will increase their fueling cost advantage even more as the price of uranium continues to rise.

From experience throughout the world with construction of reprocessing plants over a range of sizes, it seems that capital costs can be approximated by a relationship of the form shown in the following equation:

$$\frac{C}{C_{1500}} = \frac{Z^n}{1500}$$

where

C = capital cost of a plant of capacity of Z MgU/a
C_{1500} = capital cost of a plant of capacity 1500 MgU/a
n = 0.3 when $Z < 1500$
n = 0.6 when $1500 < Z < 3000$
(When $Z > 3000$, costs are approximately linear with size.)

Operating, shipping, and waste-handling costs are essentially independent of plant size. Thus, combining these factors leads to a relationship of reprocessing costs vs. plant capacity as shown in Figure 3-5. This illustrates the "optimum" size of 1500 MgU/a; below this size overall cost increases sharply due to increasing unit capital costs, while there is little economic incentive to build larger plants.

Factors that would improve the economics of reprocessing plants at regional fuel-cycle centers include:

1. Standardization of fuel designs and types. Clearly, separate plants would be designed for enriched fuel and natural uranium fuel. Even with enriched fuel, operational and shipping savings will accrue from handling standardized designs, enrichments, and burnups.

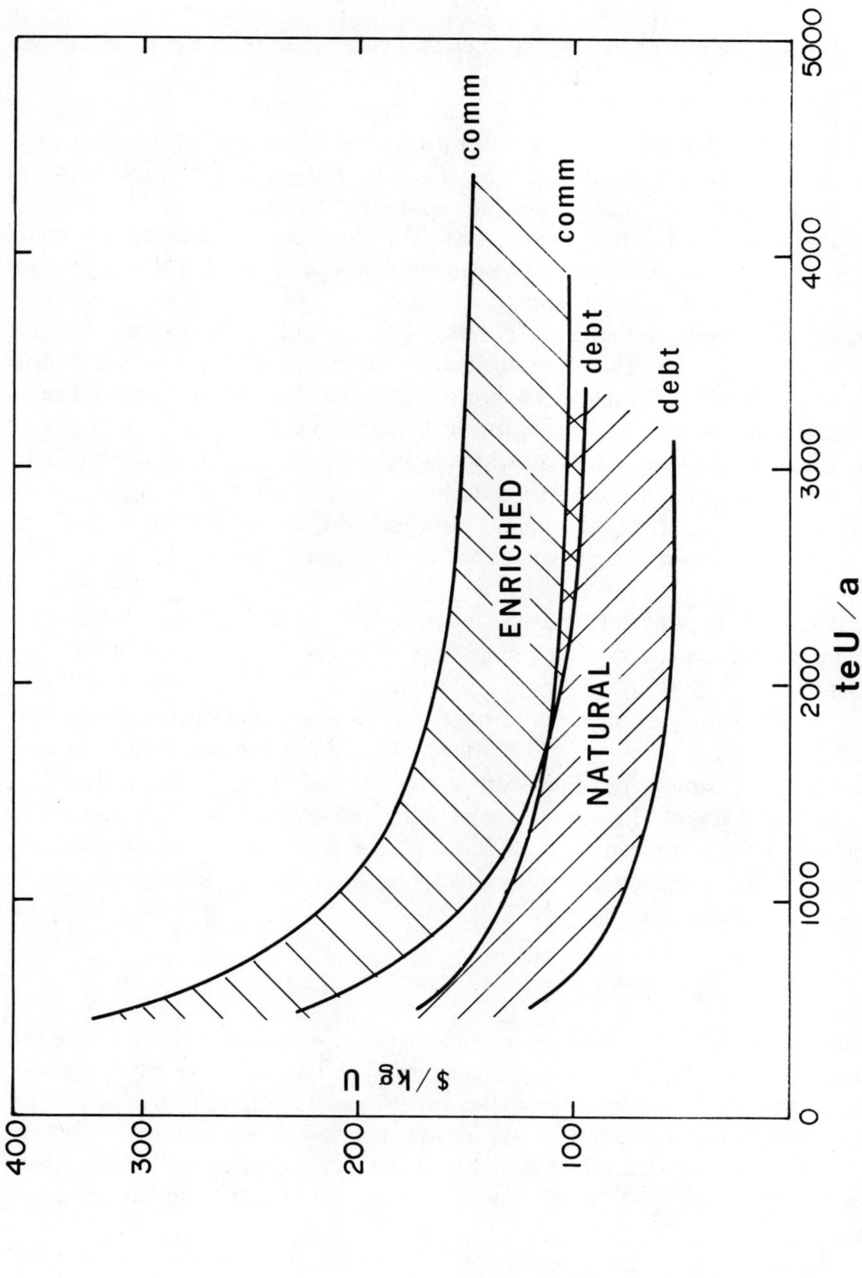

Figure 3–5. Effect of Scale on Reprocessing Costs.

2. Integration of the reprocessing and fuel-fabrication plants to reduce handling and storage costs and to minimize the possibility of diversion of separated fissile material.

On the other hand, some cost increases might arise from the following factors:

1. Delays in construction due to protracted negotiations between the joint owners will represent the probable major uncertainty in costs. For example, an extension of construction time from seven to ten years could add about 15 percent to the reprocessing cost. The exact cost increment will depend upon the shape of the capital spending curves and could realistically be in the region of 5 to 10 percent.
2. Excessive staff, training costs, hiring quotas, and so on may be expected to increase costs. These are difficult to assess accurately. However, if total staff is doubled compared to an efficient national operation, the incremental cost will not exceed 5 percent for the larger plants.
3. Quotas may be assigned to each participant for the supply of capital equipment. This could conceivably increase costs. If we assume a 20 percent increase in mechanical, piping, electrical, and instrumentation, then the reprocessing cost increase will be about 10 percent.

Overall it appears that the cost penalties due to political considerations in multinational centers would be more than offset by the advantage of increased plant size.

Since the costs are such that attracting commercial financing might be difficult, it would appear desirable to finance the plants through debt, with each participating country providing part of the required capital in proportion to its expected share of the capacity. Operating costs would then be assessed to provide a capital margin for modifications. The alternative is for the host country to finance the plant on commercial terms to provide its incentive, and charge the client countries accordingly.

CONCLUSION

The uncertainties of the cost of building and operating reprocessing plants will only be reduced by actual experience. The technical processes have been reasonably well established but the whole operation is sensitive to political considerations which may impose costly delays and revisions of agreements. The extraction cost involves also the annual charges on capital which depend markedly on the method of financing by debt or equity and the applicable taxes.

The increase of scale, from reducing the number of plants through multinational cooperation, promises cost advantages which, it must be hoped, will more than offset the costs of reaching and maintaining the necessary agreements.

REFERENCES

Crawford, R.E., and Häusserman, W. (Nuclear Energy Agency, Paris). "OECD Projected Nuclear Fuel Cycle Demand and Supply," paper presented to the IAEA Advisory Group Meeting; "Fuel Cycle Demand, Supply and Cost Trends," Vienna, November 1975.

Jonke, A.A., Levitz, N.M., and Steindler, M.J. (Argonne National Laboratory). *Nuclear Metallurgy,* Vol. 15 (CONF 690801, August 1969). AIME, New York.

Price, W.J. (executive vice president, Allied General Nuclear Services). "Reprocessing Incentives." Paper presented at AIF's "Fuel Cycle Conference '76," Phoenix, Arizona, March 1976.

Takei, M. (Institute of Energy Economics, Japan). "Nuclear Power Forecast for Japan and Asia," presented to the IAEA Advisory Group Meeting on Fuel Cycle Demand, Supply and Cost Trends," Vienna, November 1975.

Chapter 4

The Simple, Uncertain Economics of Multinational Reprocessing Centers

Mark Sharefkin

We have two objectives in this paper. The first is relatively routine: we want to identify, and bound quantitatively, the narrowly economic factors likely to matter to nations choosing between the alternatives of national and multinational spent-fuel reprocessing. The second objective is more elusive but no less important. We want an understanding of how things work in the other direction, of how, and how significantly, the choice of an institutional arrangement for reprocessing may affect the economic and technical performance of reprocessing establishments.

A word on the organization of the paper: the narrow economics of national reprocessing are presently highly uncertain, and much of our work has to be done to find sensible bounds on those costs. The section that follows attempts to bound those uncertainties with calculated fuel-cycle costs for several major alternative nuclear fuel cycles. The central question in that section is: how would a cost-minimizing firm or public servant choose between recycle and nonrecycle modes of operation? We consider three specific reactor operation modes, two of which involve reprocessing and recycle, and the third of which does not.

The first two modes would be found, side-by-side, in an economy with reprocessing and recycle. In that economy some reactors would be loaded, much as they are now, with enriched uranium fuel. Spent fuel from those reactors would be shipped to chemical reprocessing establishments, where valuable uranium and plutonium would be separated from a stream of valueless radioactive waste residuals. The residuals, suitably transformed, would be permanently disposed of. The by-product uranium and plutonium, on the other hand, can be recycled—shipped to mixed oxide fuel plants, and there fabricated into fuel elements for reactors operated in the second mode.

The third, or throwaway mode, will be the only mode found in a world in which reprocessing and recycle are, for whatever reason, ruled out. In that mode spent reactor fuel from uranium-burning reactors is a dangerous by-product, the whole of which must be disposed of. Other modes could have been considered, and may be of importance in the long run, such as natural uranium with heavy-water reactors, thorium, or others. But probably for the foreseeable future, and most likely for the period in which the structure of the international reprocessing industry will be established, the modes considered here are the important ones.

The cost information developed in the first section is required for the second section, "Incentives for Cooperation," where the more aggregative question of the economic incentives to multinational reprocessing is posed, and dollar estimates are made. The conclusions of the paper are summarized in the final section. Wherever possible, equations and other technical paraphernalia have been retrievably deposited in Appendix 4-A, and sections of that appendix are cited at appropriate places in the text.

THE ECONOMICS OF ALTERNATIVE NUCLEAR FUEL CYCLES FOR EXISTING TYPES OF LIGHT-WATER REACTORS

Why are reprocessing costs so uncertain? It is best to identify, at the outset, the sources of uncertainty in reprocessing economics.

1. The design-study basis of reprocessing cost estimates. The historical data base for spent-fuel reprocessing and mixed-oxide fuel fabrication is thin, and consequently of uncertain value in making mature industry cost projections. Although it is true that there is considerable experience with reprocessing of plutonium from "production" reactors for weapons purposes, reprocessing of spent fuel from power reactors on a commercial basis is still in the infant industry stage.
2. The importance of occupational-exposure constraints (imposed to protect the health of workers) as determinants of reprocessing cost, and the uncertain cost impact of those constraints. Large sections of chemical reprocessing plants must be fitted with costly remote-maintenance equipment. The capital cost implications of present occupational-exposure standards are uncertain.
3. Finally there are "regulatory uncertainties." A privately owned, publicly regulated chemical reprocessing industry faces the prospect of regulatory actions which are unpredictable in their substance, timing, and consequences. And those regulatory uncertainties drive a wedge between private-sector and public sector chemical reprocessing cost estimates.

While all three sources of uncertainty are significant, our purposes require a focus on number 3, for there we can see the interaction between structural arrangements and cost estimates.

Chemical reprocessing: cost estimates. We need an estimate of the cost of chemical reprocessing, in dollars per kgHM (kilogram of heavy-metal throughput), as those costs are seen by potential private entrants into the reprocessing market; similarly, we need an estimate of those costs as perceived by public-sector evaluators of proposed government investments in reprocessing capacity. Public and private cost estimates need not be the same. Private cost estimates legitimately include an allowance for the private cost of risk-bearing; but, because the government's investment portfolio is large, public investments with yields uncorrelated with national income are effectively riskless investments, and should not include an allowance for the cost of risk-bearing.

For our illustrative calculations we can mimic this private-public cost differential by adjusting the corresponding discount rates. Thus we will use public cost estimates based upon a 15 percent discount rate and private cost estimates based upon discount rates ranging from 25 to 35 percent. The latter are typical of the rates actually used by large American chemical and energy companies in evaluating reprocessing investments.

Given discount rates, and given estimates of chemical reprocessing capital and operating costs, the calculation of chemical reprocessing service prices just sufficient to cover costs is routine. Table 4-1 presents the results of three such calculations, one for each discount rate and one for each of two estimates of the initial capital cost of a 3000 tonnes/yr plant. Those capital cost estimates are relatively high, and are based upon an assumption that occupational-exposure standards will require considerable remote-handling equipment.

Economics of alternative nuclear fuel cycles. We want to compare the cost of electricity generated by reactors operating in various modes. That comparison requires two kinds of information: (1) quantities—the nature, timing, and quantities of all fuel-cycle inputs; and (2) prices (or, where prices are uncertain, price ranges).

In Table 4-2 we have assembled plausible input price ranges. For example,

Table 4-1. Estimates of the cost of chemical reprocessing, based on 3000 MT/yr (30 yr. useful life; capacity, 40% 1st yr., 60% 2d yr.)

Initial capital cost (billions of 1976 dollars)	*Target rate of return*	*Capital cost recovery charge $/kgHM*	*Total chemical reprocessing cost, including $20/kgHM operating cost*
2.0	0.15	110	130
	.25	210	230
	.35	310	330
2.5	.15	137.5	157.5
	.25	262.5	282.5
	.35	387.5	407.5

Table 4–2. Reasonable Ranges of Some Fuel-Cycle Input-Price Estimates

			With Recycle				Without Recycle		
			Low-enriched Uranium fuel		*Mixed-oxide fuel*		*Throwaway Cycle*		
		Fuel cycle Inputs	*Low*	*High*	*Low*	*High*	*Low*	*High*	*Units*
	1	U_3O_8 Ore	15	25	15	25	15	25	\$/lb.
	2	Conversion to UF_6	4	4	4	4	4	4	
	4	Uranium Enrichment	75	100	75	100	75	100	\$/SWU
	5	Fuel Fabrication	70	70	106	326	70	70	\$/kgHM
	8	Spent-Fuel Shipment	20	20	20	25	N.A.	N.A.	\$/kg
Spent-Fuel Management	9.1	Chemical Reprocessing	130	330	130	330			\$/kgHM
Spent-Fuel Management	9.2	Storage	N.A.	N.A.	N.A.	N.A.	25	25	\$/kg
Spent-Fuel Management	9.3	Shipment to Disposal	3	3	3	3	20	20	\$/kg
Spent-Fuel Management	9.4	Encapsulation	6	6	6	6	6	6	\$/kg
Spent-Fuel Management	9.5	Disposal	15	15	15	15	27	27	\$/kg

N.B.: This table, while a convenient format for a compilation of input-price information, is not a process diagram. Lest there be any confusion of the two, we emphasize that not all entries correspond to inputs required for the fuel-cycle mode associated with the column. But since in a world with recycle, end-of-cycle services are available to all customers, we have entered their price ranges in both *With Recycle* columns.

the chemical reprocessing estimates computed above become the relevant range taken for one input price—the price of chemical reprocessing services; shown in row 9.1 of Table 4-2. Other input price ranges have been somewhat arbitrarily selected. For example, the range of U_3O_8 prices used ($15 to $25) is typical of many published projections of near-term uranium prices.

The numbers labeling the rows of Table 4-2 are taken from our own more detailed listing of input quantities and timing for each electricity-producing activity. The full listing is reported in Table 4-3, which we have adapted from a recent survey report on nuclear fuels supply (Edison Electric Inst. 1976). Again, note that Table 4-3 specifies the times, measured in months from reactor core loading, at which fuel-cycle inputs must be purchased by nuclear plant managers: the length of the nuclear electric production period—almost six years from uranium ore purchase to plutonium by-product recovery—requires the systematic inclusion of working-capital charges in fuel-cycle cost calculations.

We can use Tables 4-2 and 4-3 to structure a descriptive account of some characteristics of reactor and reprocessing technologies that are central to an understanding of reprocessing economics. Note that Table 4-3 is broken into three columns: the first two are technical coefficients for the two light-water reactor (LWR) fuel cycles that will coexist in a plutonium recycle economy. In that economy, LWRs—which can burn either low-enriched uranium or mixed-oxide fuels—will be divided between the two operating modes. Plutonium will be a valued by-product of reactor operation, and the cost-minimizing behavior of reactor operators will clear the plutonium market at a plutonium price at which the costs of electricity generated by reactors operating in both the low-enriched uranium mode and the mixed-oxide mode are equal.

That equilibrium plutonium price can be determined from the technical coefficients in columns 1 and 2 alone (see Table 4-3), and is independent of the way in which operating reactors are apportioned between the two LWR modes: this result traces to our fixed-unit input assumption and to the existence of a plutonium market. Of course the equilibrium plutonium price will depend upon the prices of other fuel-cycle inputs, and we will need an explicit expression for that dependence.

In column 3 of Table 4-3 we have technical coefficients for low-enriched uranium LWR operation without recycle. In this mode of operation there is no plutonium market, and no positive plutonium price. Spent reactor-fuel elements are, after a cooling period, shipped to a waste management facility for ultimate disposal. The hypothetical ultimate disposal scheme upon which the cost estimates of the third column of Table 4-2 are based includes initial cooling, separation of the longer-lived fission products and encapsulation of those products in glass, and, finally, perpetual storage and surveillance.*

*A somewhat different, "stowaway" cycle would retain the spent fuel elements for an indefinite period, without either permanently foreclosing the reprocessing and recycle option by "ultimate" disposal or immediately exercising the reprocess-recycle option. Though we have not analyzed the economics of the stowaway cycle, the calculations are,

Now note that fuel-cycle input 1 in Table 4-3—U_3O_8 ore—is the only input which cannot be produced at constant cost: if we are driven to lower-grade uranium ores, and if improvements in extraction and concentration technologies are less than offsetting, real U_3O_8 ore prices will rise. Some care is required in incorporating those ore price increases into cost comparisons of the various reactor modes: it is important that assumed institutional constraints on reactor mode operation be stated explicitly. If it is assumed that there are no institutional constraints on reactor operation, so that reactor operators are free to make economizing choices among modes, then intermodal cost comparisons must be based upon a single ore price for all modes—the price ruling in the ore market. But if the comparison proposed is between two universes—one in which all reactors are operated in the throwaway cycle mode, and one in which recycle prevails—then cost calculations should be based upon two different ore prices, with the higher, throwaway-regime price reflecting uranium scarcity. For our purposes the single-price assumption is appropriate, since we want a measure of the purely economic determinants of the fuel-cycle choice: thus the identical ore price ranges indicated in Table 4-2.

Next, note that plutonium enters as an input only for mixed-oxide operation (input 3 in Table 4-3), and as a valued output only for recycle modes: in the throwaway mode, plutonium is a zero-price by-product. Turning to input 4 in Table 4-3, note that no uranium-enrichment service inputs are required for LWR operation on mixed-oxide fuels in the recycle regime. This is because the optimum enrichment of the uranium component of mixed-oxide fuels in present LWRs is very close to the naturally occurring fraction of U-235 in uranium, so that further enrichment is unnecessary.

Note that the fuel-cycle input—row 5 in Table 4-3—is called "fuel fabrication" (services) for all three reactor mode columns. This is somewhat misleading. Fabrication of mixed-oxide fuels entails plutonium handling, and is consequently more costly than low-enriched-uranium fuel fabrication. To be entirely faithful to our intention of enforcing a complete separation between technical and economic parameters, we could have distinguished the two kinds of fabrication services in Table 4-3. But we have, in the interests of economy of expression, chosen this sacrifice of consistency. Nevertheless, we have not neglected this difference. It shows up strikingly in the far higher cost estimates recorded in Table 4-2 for mixed-oxide fuel fabrication, ranging up to almost five times higher than the corresponding estimates for enriched uranium fuel fabrication.

The opposite choice was appropriate to, and has been made for, input 9 in Table 4-3, spent-fuel management. That row has been disaggregated in Table

---

with one caveat, very similar to the calculations for the other cycles. The caveat: the choice of the alternatives being compared is somewhat more subtle. There is of course the simple comparison of disposal and stowaway, which turns on a comparison of perpetual spent-fuel disposal and perpetual stowaway costs. But there also is the complicating feature of an uncertain future prospect of reprocess and recycle.

Table 4-3. Technical Coefficients for 1 kg Heavy Metal Charged to Reactor, Based on Fuel Burnup of 33.5 MWd/kgHM for All Cycles

		With Recycle				Without Recycle		
		Low-Enriched Uranium Fuel		*Mixed-Oxide Fuel*		*Throwaway Cycle*		
Row	Fuel-cycle Input	*Time (months)*	*Quantity*	*Time (months)*	*Quantity*	*Time (months)*	*Quantity*	Unit
1.	U_3O_8 ore	-12	16.98	-9	2.53	-12	16.98	lb.
2.	Conversion to UF_6	-9	6.50	-6	0.97	-9	6.50	kgU
3.	Plutonium	N.A.	N.A.	-6	32	N.A.	N.A.	gmf Pu
4.	Uranium Enrichment	-6	4.27	N.A.	N.A.	-6	4.27	SWU
5.	Fuel Fabrication	-3	1.00	-3	1.00	-3	1.00	kgHM
6.	Insert Fuel into Core	0	1.00	0	1.00	0	1.00	kgHM
7.	Discharge Fuel from Core to On-site Storage	41.5	1.00	41.5	1.00	41.5	1.00	kgHM
8.	Spent-Fuel Shipment	53.5	1.00	53.5	1.00	53.5	1.00	kgHM
9.	Spent-Fuel Management	53.5	1.00	53.5	1.00	53.5	1.00	kgHM
10.	Uranium Recovery	59.5	0.94 0.41% U-236	59.5	0.92 <0.2% U-235	N.A.	N.A.	kgU
11.	Plutonium Recovery	59.5	6.33	59.5	19.9	N.A.	N.A.	gmf Pu

4-2 into five spent-fuel management subrows: chemical reprocessing, storage, shipment to ultimate disposal, encapsulation, and ultimate disposal. The reason for this disaggregation is perhaps apparent: the nature of the spent-fuel management requirement, and consequently the sub-input requirements, differ sharply between the various sector modes.

Row 10 in Table 4-3 includes information on the isotopic composition of the uranium recovered at a reprocessing plant. Two special features of this component of reprocessing throughput must be considered in calculating its value. First, the concentration of U-235 is higher in by-product uranium than in natural uranium; and second, unlike natural uranium, by-product uranium contains about 0.4 percent U-236, produced from U-235 by neutron capture. These two effects are partially self-cancelling in their net effect on uranium by-product price. The larger U-235 concentration increases the value of by-product uranium while the U-236 decreases the value of the recovered uranium.

The U-236 effect on the value of recovered uranium can be traced to three distinct mechanisms, each of which decreases the economic value of by-product uranium. First, more separative work is necessary to enrich U-236-contaminated uranium to any required U-235 enrichment level. Second, in-core U-236, when subjected to neutron bombardment, gives rise to additional transuranic nuclides, some of which—Pu-238, for example—increase the shielding required in handling recycle plutonium. Third, other transuranic nuclides, along with U-236, are neutron absorbers, reducing core reactivity and therefore also reducing achievable burnup.

Calculation of an equilibrium plutonium price. We have observed that in a recycle-reprocess economy the plutonium market will clear at a plutonium price equalizing the fuel-cycle costs of electricity generated from the two recycle modes. Thus we can mimic the action of a plutonium market by treating the plutonium price as a variable to be determined by the cost-equalization condition. Because Tables 4-2 and 4-3 in effect tell us the amount, timing, and price (or price range) of all other fuel-cycle inputs, those tables are all we need to calculate our equilibrium plutonium price. And we shall see that, once we have calculated a plutonium price, we can also reproduce the revenue calculation that will be made by prospective private entrants into the chemical reprocessing industry.

First the plutonium price calculation: In addition to the price and quantity information in Tables 4-2 and 4-3, we must select a utility discount rate, and we must specify a model of the fuel cycle by dividing the variables other than plutonium price into (1) exogenous variables, which become parameters in the plutonium price equation, and (2) endogenous variables, which are retained as true variables in the price equation. We hasten to note that the utility discount rate employed by nuclear plant managers in their fuel-cycle calculations is in principle distinct from, and in practice quite different from, the other discount rates—the rates for reprocessing investment decision calculations—encountered above. For our illustrative calculations we have taken a 15 percent utility rate, roughly the pretax return on capital sufficient to guarantee a "fair" 8 percent posttax return on capital. Similarly, we have retained as dependent variables the price of U_3O_8 ore (*PUORE*), the prices of low-enriched uranium fuel and mixed-oxide fuel fabrication services (*PFABU* and *PFABMOX*), the price per unit of separative work of uranium enrichment services (*PSWU*), the price of waste disposal services (*PWASTE*), and the price of byproduct uranium (*PUREC*). (These are the symbols I use on these variables in other work I am doing on the economics of the nuclear fuel cycle.) Discounting all costs to any given point in time, equating fuel-cycle costs for the two recycle modes, and making some manipulations yields an equation for an equilibrium plutonium price. (The equation is reported in section 1 of Appendix 4-A.)

The private economics of spent-fuel reprocessing. Now we can reproduce the calculations of potential entrants into the spent-fuel reprocessing industry. We need only compare privately perceived spent-fuel costs and revenues, the revenues being those garnered from the sale of by-product plutonium and uranium.

Since 0.92 kg uranium and 19.9 gm plutonium will be recovered from each kgHM of chemical reprocessing throughput, revenue per unit throughput can be computed from these numbers and the prices for by-product uranium and plutonium. The by-product uranium price is known from the U_3O_8 price assumption and the U-236 penalty. And we have calculated an equilibrium plutonium price in terms of the ore price, the fabrication penalty, the price of reprocessing, the price of separative work for enrichment, and other parameters. In sum, we have an estimate of unit revenue. Together with our previous estimates of unit cost, we have an inequality condition for the profitability of private investments in reprocessing capacity. (This condition is recorded in section 2 Appendix 4-A.)

It will be helpful here to exhibit the results of a few illustrative numerical calculations. On the revenue side, Table 4-4 exhibits equilibrium plutonium prices and unit reprocessing revenues for two extreme sets of values of the independent variables determining plutonium price. Turning to the cost side, suppose that the 35 percent rate of return is relevant, and that the two total reprocessing cost estimates associated with that target rate of return are (subjectively) considered equally probable; these cost figures, \$330/kgHM and \$407.5/kgHM (from Table 4-1), are neatly bracketed by the two extreme unit revenue figures in Table 4-4, \$179.1/kgHM and \$599.4/kgHM.

Bringing revenue and cost ranges together, it is approximately true that a potential reprocessing industry entrant who assigns equal probability to all points in the cost and revenue ranges will conclude that net profits and losses are equally probable. This is what the often-heard comment, "The economics of chemical reprocessing are highly uncertain," should be taken to mean.

Table 4-4. Equilibrium Plutonium Price and Unit Reprocessing Revenue for Two Extreme Sets of Values of the Independent Variables

Variable	Value of Variable		Unit
	Low	*High*	
PUORE	15	25	\$/lbU_3O_8
PFABU-PFABMOX	-256	-36	\$/kgHM
PSWU	75	100	\$/SWU
PREPR	130	330	\$/kgHM
*PPU**	5.92	25.15	\$/gmf Pu
Unit Reprocessing Revenue	179.14	599.4	\$/kgHM

The public economics of chemical reprocessing. Thus far we have been talking about the private economics of chemical reprocessing, and the discount rates we have used in imitative investment-decision calculations have included the substantial risk premiums–employed by risk-averse private firms, ordinarily 10 to 20 percent above the "riskless" 15 percent rate. What of the reprocessing investment-decision problem viewed as a public investment problem, and therefore evaluated with the help of public-sector criteria?

Here we need to know how to incorporate publicly borne risk in our investment-decision calculations. No completely general rule can be given for the treatment of risky public investments. But it can be shown that, if the yield of a prospective and uncertain public investment is uncorrelated with national income, the government should be risk-neutral in its investment decisions (Arrow & Lind, 1970).

Is this risk-neutrality condition accurately descriptive of the chemical reprocessing case? Here, as elsewhere, scenarios are conceivable in which it is not–for example, a failure to invest in reprocessing capacity leads to a shutdown of nuclear generating capacity, which leads to. . . . But we believe that they are also improbable, so that risk-neutrality is the correct public investment-decision rule. This suggests the use of riskless discount rates in calculations of the public cost of chemical reprocessing, and 15 percent is perhaps a decent upper bound.

Now we can exploit the 15 percent rate of return cost figures given in Table 4-1. The result is a cost range–\$130/kgHM to \$157.5/kgHM–which sits well below the unit revenue range of Table 4-4. Illustrative calculations of this sort are slim reeds to lean on, but we have some confidence in the conclusion suggested by this particular illustrative calculation: public-sector evaluations of chemical reprocessing investments will be favorable, and even "robustly" so.

Recycle modes versus the throwaway-cycle mode. Before we rush from this conclusion to summary judgment on the outcome of a cost-benefit analysis of fuel-cycle alternatives, remember that we have been working among a somewhat restricted set of alternative reactor-operation modes. We have been (implicitly) assuming that recycle was required, either by fiat or by the comparative economics of recycle and throwaway cycles. But what if the choice between recycle and throwaway cycles is unconstrained, and left to private decision? Just as we calculated an equilibrium plutonium price (*PPU**) at which cost-minimizing reactor operators are indifferent between low-enriched uranium and mixed-oxide modes, we can exploit the technical coefficients of Table 4-3 and the price estimates of Table 4-2 to compute the discounted present cost of throwaway-mode operation. Equating this cost to the cost of recycle-mode operation at the equilibrium plutonium price, we have an equation defining the conditions under which reactor operators are indifferent between recycle modes and the throwaway mode (that equation is recorded in section 3 of Appendix 4-A). Illustrative numerical calculations of this recycle-versus-throwaway choice give the same kinds of conclusions given by our earlier calculations: entirely reasonable ranges of the exogenous variables do not decisively favor either recycle or throwaway.

INCENTIVES TO COOPERATION: SCALE ECONOMIES AND TRANSPORTATION COSTS

As noted at the outset, there is no simple, cause-effect relationship between the technical and economic data underlying chemical reprocessing economics, on the one hand, and the institutional structure of the international reprocessing industry, on the other. To the contrary, we expect that each will influence the other, with production economics delimiting the possibilities for cost savings from cooperation—subject to the assumption that, whatever the locations and scales of reprocessing plants, they are operated efficiently—and with institutional arrangements imposing constraints on efficient economic performance.

Though it is not wholly satisfactory to split these problems and treat them in turn, it is necessary to do so. Characteristically, it is easier to be somewhat precise about the first set of questions than about the second. At the heart of the first question are the familiar economic incentives to cooperation associated with production processes characterized by scale economies: where cost per unit output is lower for larger plants, the production of any given level of output from fewer larger plants rather than more smaller plants is cheaper. The larger the cost savings, the greater the incentives to cooperation, so that estimates of cost savings associated with multinational—as opposed to national—chemical reprocessing are in effect indicators of the strength of those incentives.

Central to the second set of questions are the problems and possibilities arising from multinational ownership of, multinational management of, and perhaps also multinational operating authority over, chemical reprocessing plants. Under any or all of these assumptions, must cost estimates based upon national reprocessing center data be adjusted? And if so, how? There is no clearly indicated answer to this question. There is often a presumption that multinational management is necessarily less efficient than national management, because of political constraints on staffing. Although this argument has some plausibility for a labor-intensive operation, it is of dubious relevance to a highly capital-intensive process such as chemical reprocessing. Indeed, similar constraints on employment in the capital-intensive oil industry have not been disastrous. We believe it is at least plausible that there will be no cost penalty associated with multinational management of a chemical reprocessing industry. Consequently we are free to focus upon the first set of relationships —between production economics and incentives to cooperation.

Spent-fuel reprocessing, like any chemical processing operation, is almost certainly subject to substantial economies of scale. Here, as elsewhere, scale economies imply both incentives to cooperation among parties and ways of economizing on operating-loss subsidies arising out of public, or one-party, provision. In order to make rough estimates of these effects, we need a forecast of the future size of the nuclear power industry, both domestic and worldwide. The estimates need only be order-of-magnitude accurate, and should, for our purposes, err, if at all, on the side of overestimating the size of the future industry.

For these reasons we have used a set of estimates prepared by the U.S. Atomic Energy Commission (1974). Table 4-5 presents time series derived from that source. Given industry growth projections, we can estimate the size of the possible scale-economy-related savings from multinational reprocessing. The question is: How much cheaper will it be to reprocess spent nuclear fuels in fewer larger plants than in more smaller plants? In doing this calculation we have to net out increased transportation costs arising from the probable increase in ton-mileage of spent-fuel shipments to fewer, larger reprocessing centers.

This is a generic problem with a technical name: it is a location-allocation problem for an increasing-returns production process. There is a least-cost solution to the problem, calculable with well-known formal methods. But everything we need can be gotten with simple numerical calculations, which fancier and more formal procedures merely serve to dress in respectability. So we report only some very simple, illustrative calculations. (The equally simple model underlying these calculations is described in section 4 of Appendix 4-A.)

Table 4-6, based on Table 4-5 and the additional assumption that all plants are 1500 MT/year plants, gives upper bounds on total capacity, at five-year intervals, of U.S. and foreign chemical reprocessing industries. With those upper bounds, and with the model described in Section 4 of Appendix 4-A, the estimates of annual undiscounted chemical reprocessing costs given in Table 4-7 follow easily. In that table we have compiled those costs for specific plant-size assumptions—1500 MT/year and 3000 MT/year—in order to illustrate the magnitude of scale effects. Our simplifying assumption that all plants are either 1500 MT/year plants or 3000 MT/year plants is certainly somewhat arbitrary; a

Table 4-5. Projections of U.S. and Foreign Nuclear Generating Capacity and Spent-Fuel Reprocessing and Mixed-oxide Fuel Fabrication Loads, 1975-2000

	Nuclear Generating Capacity (Thousands of Electrical Megawatts)			*Spent-Fuel Reprocessing Load (Metric Tons of Heavy Metal)*			*LWR Mixed-Oxide Fuel Fabrication Load (Metric Tons)*		
Year	*U.S.*	*Foreign*	*World*	*U.S.*	*Foreign*	*World*	*U.S.*	*Foreign*	*World*
1975	47	45	92.0	250	3100	3500	0	0	0
1980	102	140	242	1740	5000	7500	280	330	610
1985	250	387	684	4800	10,400	15,200	900	1080	1980
1990	475	780	1255	8,900	20,900	29,800	1520	2800	4320
1995	760	1367	2127	14,000	36,900	50,900	1640	3100	4470
2000	1090	2130	3220	18,300	54,400	72,700	1980	4500	6480

Table 4-6. Upper Bounds on the Expansion of the U.S. and Foreign Spent-Fuel Reprocessing Industries

Number of Spent-Fuel Reprocessing Plants of Capacity 1500 MT/yr			
Year	*U.S.*	*Foreign*	*World*
1975	1	3	4
1980	2	4	6
1985	4	7	11
1990	6	14	20
1995	10	25	35
2000	13	37	50

Table 4-7. Annual Costs of Spent-Fuel Reprocessing, 1500 MT/yr and 3000 MT/yr Cases, Based on 0.6 Scale Power Law and $200/MTHM for the 3000 MT/yr Plant

	1500 MT/yr plants		*3000 MT/yr plants*	
Year	*Number (World)*	*Annual Undiscounted Cost (billions of 1976 dollars)*	*Number (World)*	*Annual Undiscounted Cost (billions of 1976 dollars)*
1975	4	2.38	3	1.8
1980	6	2.38	3	1.8
1985	11	4.75	6	3.6
1990	20	7.92	10	6.0
1995	35	11.88	15	9.0
2000	50	20.59	26	15.6

mature chemical reprocessing industry will no doubt consist of regional centers of different sizes, with the size of any one regional center determined by the regional reprocessing load.

Finally, in Table 4-8, we have discounted the reprocessing-cost time-series of Table 4-7 to the present. The results are presented for our two illustrative plant sizes. The difference between the two discounted present-cost estimates is almost $7 billion. We again comment on the interpretation of this number: It is kind of upper bound on the monetary incentives to international cooperation in chemical reprocessing, a measure of how much can be saved, in the aggregate, with a multinational rather than an autarkic, national chemical reprocessing industry. It should be recognized that it is conceivable, though perhaps unlikely, that a private reprocessing industry, under private national or multinational control, might time capacity expansion in a scale-efficient pattern, thereby realizing these cost savings.

The reader may note that we never did net transport costs from scale-economy savings in our illustrative calculation. This is because, as indicated in Appendix 4-A, numerical estimates show that the transport-cost effect is orders of

Table 4-8. Discounted (to 1975) Costs of Spent-Fuel Reprocessing, for 1500 MT/yr and 3000 MT/yr Cases

	Discounted (to 1975) Cost of Spent-Fuel Reprocessing (billions of 1975 dollars)
1500 MT/yr	26.1
3000 MT/yr	19.8

magnitude below the scale effect, over shipment distances of thousands of miles and plant-scale differences of hundreds of metric tons per year.

CONCLUSIONS

Absolute numbers mean little in themselves: what is the scale effect incentive—of the order of billions of dollars in present value—to be compared to? Compared to either the revenues or the net profits of the electric power industry, these sums are marginal. The statement can be made stronger: Compared to the monetary value of feasible and well-understood utility-industry reforms, such as peak-load pricing, these sums are small. But this is neither a prediction that multinational reprocessing is improbable nor a suggestion that it is unimportant. The significant values at stake in issues of fuel-cycle governance are "benefit side" collective-good values. The valuation "problem" for these particular collective goods—insurance against the consequences of willful diversion, for example—make economic categories seem cumbersome and inappropriate. Here, as elsewhere, valuing collective goods is tricky business, and we should not be surprised if the issues are not settled on the "cost side."

REFERENCES

Arrow, K.J., and Lind, R.C. "Uncertainty and the Evaluation of Public Investment Decisions," *American Economic Review* 40 (1970).

Edison Electric Institute. *Nuclear Fuels Supply* (March 1976).

U.S. Atomic Energy Commission. *Nuclear Power Growth, 1974-2000.* WASH-1139, February 1974.

APPENDIX 4-A

1. Equilibrium Plutonium Price Equation

Here we simply record the equation described on p. 54-55 of the text:

$$PPU^* = 0.59\,PUORE + 0.04(PFABU - PFABMOX) + 0.16\,PSWU - 0.02\,PREPR - 0.001\,PUREC + 0.001\,PWASTE + 0.88 \qquad (1)$$

All of the variables in this equation are to be taken in the units of Table 4-4: *PPU**, for example, is in \$/gmf Pu.

Note that fabrication service prices enter the equation only in the form of the difference between (*PFABU* - *PFABMOX*), a difference often referred to as the "mixed-oxide fabrication penalty." Note also that the coefficient of *PUREC* is very small, so that the equilibrium plutonium price is quite insensitive to the price of recovered uranium. Finally, if one has some technological relation between the productivity of virgin and by-product uranium—such a relation is introduced, and used, below—then one of the two variables, *PUORE* or *PUREC*, can be eliminated. It is necessary to be careful in making sensitivity judgments from equation (1), because the magnitude of the independent variables differ considerably; but over the relevant ranges, the *PPU** is determined largely by *PUORE, PFABU* - *PFABMOX*, and *PSWU*.

2. The Condition for the Private Profitability of Reprocessing Investments

We have made the parameters on which the equilibrium plutonium price depends explicit, and the relevant equation is:

$$CREPR \leqslant 0.92\ PUREC + 19.9\ PPU^*\ (PUORE, PFAVU -PEABMOX, PREPR, PSWU) \tag{2}$$

Here *CREPR*, the privately perceived cost of chemical reprocessing per kg of heavy-metal throughput, may, but need not be, equal to *PREPR*, the price of chemical reprocessing.

3. Indifference

The indifference—between reactor operation in recycle modes and reactor operation in the throwaway mode—is

$$-16.98\ PUORE - 1.04(PFABU - PFABMOX) + 31.12\ PPU^* - 4.61\ PSU + 0.51\ PREPR - 0.48\ PUREC + 29.12 = 0 \tag{3}$$

If the left-hand side of (3) is greater than zero, the throwaway mode will be chosen by a cost-minimizing reactor operator. If it is less than zero, one of the recycle modes will be chosen.

4. Scale-Economy Savings

First note that, from the data in Tables 4-5 and 4-6, we can extract a rough estimate of the aggregate scale-effect cost savings from operation of plants of 3000 MT/year instead of 1500 MT/year. Thus:

$N(t)$ = number of plants of capacity Q_1 in year t; $t = 1, \ldots, T$

$TC(Q_1)$ = total cost (per unit throughput) of one-year's operation, i.e., of throughput Q_1 for one year

r = discount rate

The discounted present cost of reprocessing the spent-fuel load $Q_1 N(t)$ over the years $t = 1, \ldots, T$ is:

$$DPC_1 = \sum_t N(t) TC(Q_1) \exp. (-rt) \tag{4}$$

Similarly, the discounted present cost of reprocessing the same spent-fuel stream, but with plants of capacity $Q_2 = (Q_1/2)$, is:

$$DPC_2 = \sum_t 2N(t) TC(Q_2) \exp (-rt) \tag{5}$$

Suppose we assume that a simple power law is accurately descriptive of the scale economies available in spent fuel reprocessing. Laws of this kind are amply grounded in the facts of geometry and in experience, with a 0.6 power law generally used. This 0.6 law may understate chemical reprocessing scale economies, since those economies originate not only in geometry but also in the indivisibilities associated with large investments in occupational-exposure shielding and in remote-handling equipment. Then:

$$TC(Q_2) = (Q_2/Q_1)^{0.6} TC(Q_1) \tag{6}$$

Returning to our particular case in which $(Q_2/Q_1) = 1/2$, we have:

$$DPC_2 = (2)^{.4} DPC_1 \tag{7}$$

This relationship, which is independent of the discount rate chosen, says that the discounted present costs of meeting reprocessing load with plants of 1500 MT/yr instead of plants of 3000 MT/yr will be higher by a multiplicative factor of $(2)^{0.4} = 1.32$. All that is necessary now is a crude estimate of DPC_1. Returning to Table 4-1, such an estimate can be constructed by taking one, or a range, of values from the ranges of chemical reprocessing costs estimated, computing the corresponding value or range of values of current costs corresponding to the fuel reprocessing load scenarios upon which Table 4-6 is based, and then discounting to the present at an appropriate social discount rate. The results of a set of such calculations—15 percent has been used—are compiled in Tables 4-7 and 4-8, and provide a range of estimates of in-plant spent-fuel reprocessing costs.

We can, I think, convince ourselves very quickly that these differentials are orders of magnitude larger than the transportation-cost effects associated with the longer shipments of spent fuel to fewer, larger-scale plants, so that a naive cost-minimizing solution will have reprocessing load served by plants of the size at which scale economies are exhausted. As we have emphasized above, the total transport cost is properly calculated as the bill, at the optimum, in a cost-minimizing location-allocation problem. But because the reprocessing costs dwarf transport costs over the relevant range, we can escape here without doing that calculation. To persuade ourselves that the relative magnitudes of the in-plant reprocessing and transportation costs are as claimed, it is easiest to construct an extremely simple-minded model of a world in which spent-fuel waste loads are generated uniformly over an area A at a rate W/A tonnes per year, where W is the total annual waste load. If there are N plants, and transport costs per mile per metric ton are t, then total transport costs TTC are given by

$$TTC = \frac{t\sqrt{A}}{3\sqrt{\pi}} \qquad \frac{W}{\sqrt{N}} \tag{9}$$

From Equation (8), two time series, corresponding to worlds with plant scales 1500MT/year and 3000MT/year, can be generated, convenient fictions helpful only in providing a sense of the magnitude of the transport cost effect.

Chapter 5

Safeguarding Reprocessing Facilities: The Impact of Multinationalization

Lawrence Scheinman*

Each stage of the nuclear fuel cycle presents its own proliferation and safeguards problem. Certain stages pose greater proliferation risks than others because they produce or utilize weapons-usable special nuclear material. Chemical reprocessing facilities are a case in point. They are particularly sensitive because they make plutonium available through its separation from irradiated fuel assemblies. Plutonium is not the only weapons-usable material derived from the nuclear fuel cycle; but, despite its highly radioactive and toxic characteristics, it is an especially attractive target material because of its potentially greater availability, easier access, and lower technological and resource requirements than the principal alternative, highly enriched uranium.

The high price and political vulnerability of oil has stimulated increased interest in nuclear power in many nations. With the advent of large-scale nuclear power programs in an expanding number of nonnuclear weapons states, the possibility of widespread increased availability of separated plutonium has become a more acute problem. Many states interested in nuclear power reactors are also interested in closing the nuclear fuel cycle, particularly through the construction and operation of reprocessing facilities. The incipient trend toward the proliferation of reprocessing capability has intensified concern over the effectiveness of safeguards designed to detect unauthorized diversions of material from reprocessing plants.

Interest in acquiring a reprocessing capability stems from a number of considerations, most of which have nothing to do with a desire to maintain or pursue a nuclear weapons option. Rather, the interest derives from economic,

*The views expressed in this paper are the author's alone and do not necessarily reflect the official views or policy of ERDA.

political, and technological considerations and the objective of reducing the degree of dependence on outside sources for basic energy requirements. Once having made a commitment to increase the role of nuclear power in their national energy systems, states must consider the question of certainty of supply of power-reactor fuels, the political implications of exclusive reliance on foreign suppliers, and the impact of fuel imports on their economic situations.

The extent to which either the United States or alternate suppliers, such as Urenco, Eurodiff, or the Soviet Union, will be able to ensure future supplies of low-enriched uranium adequate to meet the needs of light-water reactors (LWR's) remains uncertain. It is generally accepted that the recycling of separated plutonium in the form of mixed oxide fuel, and the re-enriching of depleted uranium recovered through reprocessing spent fuels could add about 25 percent to the nuclear fuel supply.* No commercial reprocessing facility capable of receiving and reprocessing high-burnup LWR fuels is in operation today in any of the traditional nuclear supply states. A tripartite consortium, United Reprocessors, involving France, West Germany, and Great Britain, does exist, and Allied General Nuclear Services is constructing a facility in Barnwell, South Carolina; but none is currently operational. Nations committed to augmenting the role of nuclear power in their energy economy therefore may regard the alternative of developing national reprocessing capability as prudent management of their long-term nuclear fuel needs and as enhancing the goal of reduced dependence on external supplies.

Technological reasons also contribute to an explanation of interest in developing reprocessing capacity. A number of countries, particularly those with high energy consumption and substantial electrical power grids (that is, advanced industrial states) are interested in investigating the opportunities presented by breeder technology. They may seek a reprocessing capability partly to conduct research in this area, perhaps with a view to becoming suppliers themselves, and partly to develop a plutonium stockpile as a basis for an eventual breeder-based nuclear economy.

Finally, reasons of politics and prestige may dictate national decisions to close the back (and front) end of the nuclear fuel cycle. Nations that are parties to the Non-Proliferation Treaty (NPT) regard the provisions of Article IV of that treaty as permitting access to all peaceful nuclear technologies and as a basis for legitimately pursuing the goal of full fuel-cycle technology. In summary, interest in chemical reprocessing facilities remains high, and there is some likelihood of the expansion of their number and location, compelling us to pay close attention to the potential problems that the development of reprocessing activities may create for realization of nonproliferation objectives.

*The added amount derived from reprocessing also may be available through other means such as lowering the tails assay of enriched uranium and increasing enrichment capacity.

ACCESSIBILITY OF TECHNOLOGY

Safeguards issues can be dealt with in legal, political, and technical terms, although technical and political considerations are highly interdependent and frequently overlap. Whether or not a chemical reprocessing plant, for example, is "safeguardable" (that is, whether a diversion to unauthorized purposes could be detected) is both a technical and a political question. Technical criteria can be generated to serve as a basis for determining "safeguardability," but political considerations can enter at two points: (1) in determining what constitutes adequate criteria and which criteria should be applied; and (2) whether the application of those criteria or the method of evaluating data and information derived from their application is adequate. The central issue can be summarized in the following way: Whether it is possible to define and apply technical criteria for safeguardability, which simultaneously are *acceptable* to the party to whom the safeguards are applied and *credible* to any third party.

Before proceeding further, it would be useful to explore briefly the means by which a nation can acquire a reprocessing capability. Because of our greater familiarity with the limitations, conditions, and controls that would apply if a state were to seek material or technological assistance for reprocessing purposes from the United States, we shall base our remarks on the U.S. case. For reasons that will become evident, we believe that the situation would not be substantially different for most other nuclear suppliers.

Technical information relating to chemical reprocessing is largely available in the open literature. The United States declassified a great deal of fuel-cycle technology in the context of the Atoms for Peace program including information that related to reprocessing. So a nation interested in acquiring basic technological information on chemical reprocessing would have little difficulty. Technological information in the form of engineering know-how (design and construction) is another matter, however. Prior to 1972 it was theoretically possible for U.S. citizens or firms to assist in the design or construction of reprocessing facilities even in countries that did not have formal nuclear Agreements for Cooperation with the United States or did not accept safeguards. The Atomic Energy Commission (now Energy Resource and Development Administration, or ERDA) regulations were revised in 1972 to require specific authorization by the ERDA administrator for any person who was subject to the jurisdiction of the United States to engage directly or indirectly in chemical reprocessing activities outside the United States (10 CFR § 810). Whether or not such an authorization will be granted depends on whether the activity would be inimical to U.S. interests. In reaching that judgment factors such as whether the recipient nation is an NPT party, or would accept safeguards on the proposed activity, or has a formal Agreement for Cooperation with the United States are taken into consideration. Granted the relatively open access to technical information on reprocessing technology, one should not underestimate the antiproliferation effect of limitations on the transfer of engineering know-how and industrial design. While U.S. industry still has not put a commercial reprocessing plant into operation, it has built and operated plants for government purposes which have operated with high efficiency over long periods of time.

At a minimum the aforementioned provisions foreclose the unregulated export of sensitive technologies from the United States. In addition, as a matter of policy the United States has adopted a practice of restraint toward the supply of sensitive exports, including enrichment and heavy-water production as well as reprocessing, and of active encouragement of multinational facilities or at least supplier involvement in their transfer. It has sought to encourage other suppliers to adopt similar policies of restraint and encouragement with some measure of success. Implementation of these policies, however, remains purely voluntary.

When technology is exported, current U.S. policy is to require assurances that any sensitive facilities built using transferred technology (that is, any replication) will be placed under international safeguards. This concurs with provisions incorporated in the recently concluded agreements between France and Pakistan and between the Federal Republic of Germany and Brazil, respectively. Finally, nuclear agreements entered into by the United States include provisions for U.S. approval of the location and undertaking of reprocessing of U.S.-supplied materials and of fuels irradiated in U.S.-supplied reactors. In some instances agreements also include an option for the United States to purchase special nuclear material produced by reactor operation.

The export of special nuclear material, reactors, and other production facilities such as enrichment plants or utilization facilities (i.e. equipment capable of using special nuclear material) from the United States can only take place pursuant to an Agreement for Cooperation between the United States and the recipient state. These agreements include guarantees that supplied material and equipment will not be used for nuclear weapons or any other military purpose, and they provide for verification safeguards. Insofar as reprocessing equipment is concerned, since much of it is general-purpose equipment that might be put to a variety of uses unrelated to nuclear activities, there are only limited controls over access. Only the export of specially designed equipment requires a validated (that is, specific) license issued by the Department of Commerce after consultation with ERDA. Among the few reprocessing-related items subject to these controls are fuel-chopping equipment, counter-current solvent extractors, and certain specially designed tanks and process-control equipment. Depending on the characteristics of the material to be reprocessed, some or all of these items might be dispensed with.

As in the case of technology transfer, the United States has sought to achieve common export guidelines for the supply of nuclear material and equipment especially designed or prepared for the processing, use, or production of special fissionable material. Several years ago, in the implementation of Article III(2) of the NPT (requiring parties to the treaty not to supply equipment or material especially designed or prepared for the processing, use, or production of special fissionable material to any nonnuclear-weapons state unless placed under required safeguards), a list of items whose supply would trigger International Atomic Energy Agency (IAEA) safeguards was developed through the joint action of a large number of supplier states. Reprocessing plants as well as equipment especially designed for such plants were included on that list. More recently a number of supplier states have reaffirmed their commitment to require IAEA safeguards in the transfer of any sensitive facilities, equipment, or—as we have seen—technology.

This brief review of the nature and evolution of efforts to control the transfer of sensitive technologies and equipment such as chemical reprocessing, confirms that the problems associated with widespread reprocessing capability are recognized and, at least to some extent, being coped with. But what these limited equipment and the more stringent technology controls also reveal is that there really are no *major* technical or material obstacles to prevent a nation of even modest technological means from developing a small reprocessing facility for weapons purposes.

PROBLEMS OF SAFEGUARDS

In view of the above conclusions concerning the accessibility of reprocessing technology or equipment, the importance of defining and applying adequate and effective safeguards at this point of the nuclear fuel cycle becomes clear. Heretofore, reprocessing plants for commercial grade fuels have existed in only a few countries, and even then only on a pilot-plant level. While there are no commercial plants currently in operation, several sizable facilities are in various stages of completion in nuclear-weapons states (such as, the Windscale conversion in Great Britain, La Hague in France, and Barnwell in the United States) and, more importantly, a number of small-scale plants have been or are being constructed in non-nuclear-weapons states (India and Japan, for example).

IAEA's safeguarding experience in reprocessing has been limited to experience and development exercises at Nuclear Fuel Services (West Valley, New York) and participation in some Euratom planning exercises for safeguarding at Eurochemic (Mol, Belgium). The main current activities involve the development of standards, procedures, and technical criteria to be applied to operating reprocessing facilities. This is an exceptionally important period, for the determination of technical adequacy undoubtedly will reflect political evaluations concerning levels and speed of detection of possible diversions; and what is established as baseline acceptability today will be precedent-setting for future safeguards agreements and standards. Today, however, there is not, at the national or international level, any firm or experiential basis upon which to make confident judgments concerning the actual safeguardability of reprocessing facilities.

What is clear is that reprocessing facilities present significant safeguards problems because of the numerous opportunities that exist for diversion either at the input (head) end of the plant or at the output (back) end, and because of the general difficulty in obtaining a highly reliable physical inventory. Diversion could occur as a result of withdrawing fuel assemblies from storage canals and holding them for clandestine reprocessing, or from introducing supplementary irradiated material into the dissolver tanks; it could also occur after adding piping to draw off the product following dissolution or after separation of the plutonium and uranium streams, or at the product end of the cycle itself.

Inspector-General Rudolf Rometsch of IAEA, in discussing material accountacy problems, noted that as the character of the safeguarded material changes, and as throughput volume increases, larger absolute uncertainties will occur concerning material unaccounted for, because accuracies in a material balance closing are a fairly constant percentage of the throughput (U.S. Senate, 1976).

These uncertainties can quickly transcend the threshold amounts of material needed to make a nuclear explosive device. Under these circumstances the nature of safeguarding techniques—and in particular the role of surveillance and containment procedures as well as the frequency and duration of human inspection—take on increased significance.

The IAEA has the necessary legal basis to apply safeguards to reprocessing plants. Article XII of its Statute gives the agency the right to apply safeguards to the extent relevant to the arrangement in question whenever requested to do so by the parties. Two general documents, INFCIRC/153 and INFCIRC/66/Rev.2, which apply respectively to NPT and to non-NPT parties, give substance to the statutory mandate.

Prima facie, both INFCIRCs give the IAEA all necessary rights to apply adequate safeguards to reprocessing plants. INFCIRC/66/Rev.2, for example, provides in Annex I that "a reprocessing plant having an annual throughput exceeding 5 effective kilograms of nuclear material, and the safeguarded nuclear material in it, may be inspected at all times." It is understood that for plants having an annual throughput of more than 60 effective kg, the right of access at all times would *normally* be implemented by means of continuous inspection.

INFCIRC/153, paragraph 80, provides that the maximum routine inspection effort for facilities involving plutonium shall be determined by the formula: $30 \times \sqrt{E}$, where E = annual throughput of nuclear material expressed in effective kg, which would seem to suggest continuous inspection for even a rather modest sized plant (e.g., a 20 to 30 MTU/yr plant capable of producing approximately 120 kg Pu/yr, for example).

These provisions, however, are not self-executing. They must be implemented through the negotiation of safeguards agreements and subsidiary arrangements between the IAEA and the state or states to whom the safeguards are to be applied. The safeguards agreements themselves are basically standard, but the subsidiary arrangements are not. They are detailed statements of the manner in which a safeguards system will be applied in a particular jurisdiction. While the IAEA places considerable emphasis on the value of standardizing procedures, it faces two potential problems in doing so.*

On the one hand, two safeguards regimes exist that differ from each other in specific ways and that must be implemented in accordance with their prescribed purpose and within their established limits. As Benjamin Sanders has pointed out, in contrast to INFCIRC/153, INFCIRC/66/Rev.2, does not contain specific technical objectives for safeguards or specify conclusions to be drawn from safeguarding activities by the IAEA (Sanders, 1975). (*CRS*, U.S. Senate Congressional Research Service, 1976, p. 96). Furthermore, it is apparently not absolutely clear to what extent that document permits the application of containment or surveillance techniques to complement materials accountancy. Those techniques were developed largely to meet IAEA responsibilities under INFCIRC/153, which emphasizes instrumentation and technology over physical presence. States negotiating safeguards arrangements for reprocessing facilities under INFCIRC/66/Rev.2 might quite legitimately demur to proposals that they act as if they were NPT parties negotiating under INFCIRC/153 and thereby accept whatever methods and procedures were normative in the latter case.

On the other hand, even with respect to the same system, different states or groups of states may negotiate different arrangements. Differences would arise less with respect to the category of activity concerned (such as specifying the various materials-balance areas or identifying the accounting procedures to be followed) than with respect to the manner in which the safeguards were in fact to be carried out (that is, identification of particular strategic points for the measurement of nuclear material or the installation of surveillance equipment; whether the latter would be used; and the scope and duration of a routine inspection). The Euratom-IAEA agreement, which provides for IAEA verification of Euratom safeguards, is a striking example of de facto and perceived differences in the application of international safeguards and of concern over discriminatory treatment on the part of some non-Euratom states (Willrich, 1973). In both this situation and in the case of application of different baseline documents, the differences could place strains on a judgment that a particular safeguarding system is in fact adequate to detect unauthorized diversions on a timely basis.

*The following discussion is not intended to suggest that problems will arise in every instance but only that they may occur in some cases and thereby impact on safeguards implementation.

There is sufficient ambiguity in both baseline safeguards documents to raise questions concerning safeguards adequacy when dealing with as sensitive a facility as a reprocessing plant. Nowhere, for example, is it stated just how large a diversion should be detected, or how quickly it should be detectable. The absence of a technical standard can lead to reduced confidence by third parties in the application of safeguards to a particular facility or in a particular jurisdiction. In addition—and of even greater importance—is the fact that under the IAEA systems, the precise safeguarding arrangements (that is, scope, duration, character, intensity, and so on) constitute privileged information to which no one other than the IAEA or nation in question has access. Similarly, the findings of IAEA safeguarding under INFCIRC/153 are unavailable to any third party, and the latter might be unable to judge to its satisfaction the effectiveness of the safeguards being applied. Secrecy of information squarely raises the question of whether safeguards can be simultaneously acceptable to the party to whom they are applied and credible to a third party. If the latter objective cannot be satisfied, then the value of safeguards is in doubt. The IAEA has recognized this as a potentially serious problem and apparently plans to take steps to better inform its board of governors concerning safeguards implementation. But it must find a way to do so that does not violate its obligations to individual states. Whether this can be done and third parties still be satisfied remains to be seen.

An additional problem which, in certain respects, transcends the other issues is whether, in the case of a highly sensitive facility with large quantities of weapons-usable materials such as a reprocessing plant, the adequacy of safeguards does not ultimately depend on the continuous presence of international inspectors. The technical limitations on other safeguards techniques, the high risk of diversion, and the presence of substantial quantities of stored separated plutonium are persuasive reasons for maximizing access to such facilities by IAEA inspectors. Physical presence, or at the very least a high frequency of inspections, may be essential if there is to be timely detection of unauthorized diversion (that is, detection in time to do something). The application of locks and seals on plutonium storage units or of surveillance equipment during a presumed clean-up period at the plant will not be worth a great deal unless IAEA inspectors are ensured unlimited and instantaneous access to determine whether any tampering or illicit conduct has occurred. As noted above, however, it is not *a priori* certain that either of the basic safeguards systems *ensures* that the IAEA can undertake continuous inspection in every case where it determines such inspection is necessary. Indeed, it is quite uncertain that there is a clear, unambiguous, and agreed-to understanding of just what continuous inspection involves.

To summarize, reprocessing plants (along with fuel-fabrication and enrichment facilities) create considerably greater safeguards problems than other fuel-cycle components. A legal basis to apply comprehensive safeguards on national facilities exists, but technical criteria for "adequate" safeguards are still being developed; their definition will entail political choices that may set a precedent for future adequacy of safeguards. Certain features of the international safeguarding system, particularly protection of "proprietary" information, the need for compromise in order to secure agreement in the first place, and resultant pressures toward lowest common denominator criteria or solutions, make informed judgment on the safeguardability of such facilities problematic.

MULTINATIONAL OWNERSHIP OR CONTROL

In the previous section we tried to identify some of the problems of safeguarding reprocessing facilities. Among them was the high risk of diversion. Also examined was the problem of credibility raised first by the limited flow of information about safeguards implementation and second by the fact that in a large facility, the acceptable material unaccounted for (MUF) in percentage terms will exceed in absolute terms a critical mass for nuclear weapons. The importance in this context of continuous inspection was also discussed.

There remains the question of whether, from a safeguards point of view, multinational ownership or control of reprocessing facilities offers any special advantages. In attempting to reach a preliminary judgment we shall: (1) set aside the question of institutional and financial complexities associated with the structuring of multinational enterprises; (2) recognize that the driving force behind a national decision to acquire reprocessing capability may have little or nothing to do with economic considerations; and (3) make no assumption about the validity of the economies-of-scale argument. All three of these considerations are treated elsewhere in this book.

A number of reasons can be advanced in support of the proposition that multinational facilities enhance safeguardability of reprocessing facilities. Their order of presentation does not necessarily reflect their relative importance.

1. There can be little doubt as to the value of multinational facilities in reducing the risk of diversion by virtue of the participation of several nationalities, each watching the other. Euratom is a major example of an adversary process built into a cooperative venture. Few observers reject the argument that having Belgians and Frenchmen watching Germans, and Germans and Hollanders watching Italians is better than any one nationality watching itself, even with external verification as in the case of bilateral or IAEA safeguards. Moreover, the political objective of European integration that characterized Euratom would not have a counterpart in a commercially grounded multinational venture, thus reducing concern that "higher" interests would subvert the adversary process. In the crudest terms, it is easier to accept a one percent

MUF in a large-scale multinational reprocessing plant than it is if the plant is purely national. The risk of diversion would appear to be somewhat reduced by virtue of each partner participating in monitoring.

2. Multinational facilities offer increased protection against nationalization of a plant by the host country. To be sure, the seizure and nationalization of a multinational plant cannot be completely discounted, but that risk would appear to be inverse to the number of parties involved. Multinational control adds more legal and political obstacles to the renunciation of commitments.

3. Associated with the previous point, the fact is that with multinational ownership or control the risk of international inspectors being denied access to facilities or thrown out of the country is reduced.

4. The implementation of multinational facilities would very likely reduce the number of reprocessing plants to be placed under international safeguards, thereby alleviating for the IAEA a potential manpower and financial burden. The widespread development of safeguarded reprocessing facilities would seriously tax the IAEA. Assuming that a modest-sized plant were to be placed under continuous inspection, the minimum number of personnel that could be assigned to that facility would be three. Taking into account rotation, vacations, and support personnel, as many as nine persons might have to be earmarked for such a plant. Multinational facilities could clearly contribute to a more efficient allocation of resources *if they are substitutes for, rather than additions to, national reprocessing capability.*

5. In view of the fact that in negotiating safeguards agreements and subsidiary arrangements the IAEA often may be in a position of having to compromise, and that awareness of this can affect the perceived credibility of the application of safeguards, another advantage of multinational ownership or control is that it may enhance safeguards credibility. A multinational board of directors that included representatives from states known to be strongly supportive of international safeguards could be counted on, at a minimum, not to weaken the safeguards system and very possibly to upgrade its application to the facility in question. No general rule can be offered in this regard; it is largely a question of who the participants are. A supplier state such as the United States participating in such an arrangement would have added opportunity to evaluate how its facility partners were using the fuels supplied.

6. If regional in character, a multinational facility could help reduce suspicions in countries about their participating neighbors' nuclear weapons intentions, thereby contributing both to nonproliferation goals and to increasing regional political stability. Since international political considerations and evaluations of national security interests are major factors in national decisions about whether or not to acquire a nuclear weapons option by closing the fuel cycle, this value should not be underestimated.

7. A final important advantage of multinational fuel-cycle centers is that if they involve co-location of reprocessing and mixed oxide fuel fabrication facil-

ities, problems of safeguarding separate fuel fabrication or plutonium-storage facilities and of threats to the transport link in the fuel cycle are reduced. No nonproliferation purpose would be served by a facility that merely reprocessed spent fuel and returned pure plutonium to a non-nuclear weapons state. A number of options for dealing with separated plutonium exist; one of the most attractive would be the on-site fabrication of fuel elements and total elimination of the shipment of plutonium in any weapons-usable form. Co-location is not peculiar to the multinational concept. It could apply as well in binational or even purely national contexts. In the latter event, however, only the physical security aspect of safeguards would be covered; there would be no assurance against governmentally inspired diversion.

A number of cautionary notes about the nonproliferation value of multinational reprocessing facilities merit discussion. It is important to bear in mind that the development of multinational fuel-cycle centers does not preclude the construction of small-scale reprocessing facilities even by participating members in the multinational enterprise. The costs and length of time involved in doing so might be increased if a state were to have no external assistance in building a plant; but, as emphasized at the beginning, there is no insuperable technological or material barrier to the construction of reprocessing facilities. Large plants entail sophisticated engineering know-how; small facilities, especially ones built to handle low-burnup fuel (such as those from research reactors or from low-exposure heavy-water reactors), pose fewer problems. In view of the frequent assertion that reprocessing plants are very costly, one should remember that, measured against the capital and operating costs of a nuclear power program, the cost of a domestic reprocessing plant of modest scale may be quite acceptable and very much within the means of a state that can afford a power program in the first place. The only thing that would be required is a political decision to proceed. So multinational centers are not preventives, although they may have an impact on a state whose only concern is economic and that is convinced that economies of scale support the multinational concept. At the very least the existence of and access to a multinational facility (or even to alternate services by a supplier state) reinforce the case of those opposed to national reprocessing.

A second point relates to technology transfer. If participating states in a multinational facility are involved in the design, construction, and operation phases of the plant rather than only in management decisions or merely sharing the cost of the facility in exchange for guaranteed reprocessing services, they gain access to the know-how and engineering aspects of reprocessing. This raises the problem of the proliferation of technology that is difficult to safeguard. One may argue that the technology transfer involved is less than it would be if a supplier state were to enter a binational agreement to construct a reprocessing facility; but this is a weak response. One solution to this problem might be to adapt to multinational centers the formula utilized in the trilateral agreements between IAEA, Brazil, and West Germany as well as between IAEA, France, and Pakistan, wherein conclusive presumptions of replication attach to any facility built within a specified time period, thus requiring notification and acceptance of IAEA safeguards on the new facility. While this is not a foolproof solution, no solution is. Yet this might serve a useful controlled-proliferation purpose even if it does not eliminate entirely the possibility of proliferation. And compared to the risks involved in the spread of national reprocessing facilities to a large number of nations, granting access to a certain range of technology entails relatively limited costs.

A third problem is what we might call the Euratom syndrome. We refer here to the fact that one of the more difficult problems that has arisen in the context of international safeguards negotiations is that of the similarity or differentiation of treatment accorded the group of states as distinguished from the single state. IAEA has undertaken to "verify" Euratom safeguards, although it remains to be seen what, in practice, this concept of verification is all about. There is the possibility that the management of a multinational enterprise would contend that a multinational facility is no different from Euratom and should be accorded similar if not identical treatment insofar as the application and implementation of safeguards is concerned. Claims of this nature would resurrect the problems of equity and nondiscrimination that permeated earlier IAEA negotiations. While it is not predictable that this type of problem will arise, its possibility should not be discounted. On the other hand, it is entirely possible that a multinational facility unburdened by sovereign sensitivities might be quite receptive to IAEA safeguards, thus easing Agency tasks and offering added opportunities to test and evaluate a variety of safeguards procedures.

While additional arguments for and against multinational facilities can be offered, the foregoing raises the central problems and discusses the main themes. On balance it would appear that safeguard-ability is enhanced by collective involvement and responsibility. What remains unanswered, however, is the broader question of whether multinational control or ownership serves non-proliferation objectives enough so that the effort to structure such centers is worth the political and other costs it may entail. Multinational facilities are intended as substitutes for, not as supplements to, national reprocessing plants. Participants must be persuaded to forego national closure of the fuel cycle despite the fact that they may be parties to the NPT and that the principal supplier states, including some non-nuclear weapons states, may themselves have achieved national closure. In this situation of de facto inequality the participants will have to perceive substantial benefits and limited political costs in joining a multinational enterprise. If they view participation as only of marginal value, other concessions may have to be made by the proponents to secure acquiescence of the other nations. There is a point at which cost begins to outweigh benefits; recognizing that point and averting a situation in which the value of multinationalization is outweighed by the cost of its implementation will be the challenge of the near-term future.

REFERENCES

See Code of Federal Regulations 10 CFR 810.

Sanders, Benjamin. *Safeguards Against Nuclear Proliferation*. An SIPRI study. Stockholm, 1975.

Scheinman, Lawrence. "Political Implications of Safeguards," in Mason Willrich, ed., *International Safeguards and Nuclear Industry* (Baltimore: Johns Hopkins Press, 1973), ch. 9.

U.S. Senate Congressional Research Service. *Nuclear Weapons Proliferation and the International Atomic Energy Agency: An Analytical Report Prepared for the Committee on Government Operations*. March 1976 (cited there).

Chapter 6

The Impact of International Safeguards Arrangements Upon Nuclear-Fuel-Cycle Operation

C. Beets*

FOUNDATION OF SAFEGUARDS

From the outset of military as well as peaceful applications of nuclear energy, both governmental and industrial authorities have recognized the imperative requirement to establish a system of safeguards against possible misuse of this power by a state or by a clandestine organization. Abraham Ribicoff and Charles Percy in an analysis of the problem published in April 1975, stated:

> It is fast becoming clear that, as nations acquire civilian nuclear technology to attain energy self-sufficiency, they also acquire the materials, personnel and ability to build nuclear weapons. . . .
>
> Of no less concern is the fact that the rapid increase in nuclear technology and materials throughout the world coincides with the world-wide problem of brazen, sophisticated terrorism. Preventing the theft of weapons-grade materials by subnational groups has become as serious a problem as detecting the diversion of such materials by national governments (Ribicoff & Percy, 1975).

The authorities of all countries are becoming more keenly aware of the importance of what is at stake, and are paying increased attention to safeguards and to the protection of nuclear materials and installations. Authorized opinions

*My sincere thanks are due to Dr. Detilleux, general manager at Eurochemic, who permitted this work to be carried out. I also express gratitude for the contribution of Dr. van der Stijl of the Euratom inspectorate, who brought in his valuable experience and for the discussion with Dr. Bemelmans and Dr. Challe of my staff, concerning the presentation of this paper.

seldom go as far as to recommend the abolition of nuclear energy; yet numerous voices clamor for a strengthening—or at least for a strict application—of existing protection systems (USACDA, 1974; Barnaby, 1975). Such a strengthening depends on the optimization not only of the technologies but also of the institutional mechanisms (Willrich & Taylor, 1974). Clearly the decision is of a political nature.

THE PARTITION OF THE RISKS BETWEEN SAFEGUARDS SYSTEMS AND PHYSICAL PROTECTION SYSTEMS

The problem of nuclear material safeguards is clearly a worldwide one. For this reason international systems had to be established, notably those of the International Atomic Energy Agency (IAEA) (1955) and of Euratom (1957). These systems all rest upon the basic undertaking, on the part of every contracting state, that nuclear energy shall not be diverted from its peaceful applications.

If a state accepts safeguards, it has to make sure that the nuclear installations—or, more generally, the citizens under its jurisdiction—fulfill the corresponding obligations. In particular, any failure to comply with these obligations must involve the application of sanctions at the appropriate level. To these ends, the state has to promulgate a set of regulations and, on the other hand, establish or accept one or more control authorities endowed with the necessary degree of independence to enforce the regulations. The authority or authorities are to verify that no material or equipment has been used in activities that infringe on the basic undertaking.

The Safeguards Committee of the IAEA has taken great care in defining the "Structure and Contents of Agreements between the Agency and States required in connection with the Treaty on the Non-Proliferation of Nuclear Weapons" (IAEA, 1971a), wherein the objective of safeguards is clearly re-stated:

> *Art. 28:* The Agreement should provide that the objective of safeguards is the timely detection of diversion of significant quantities of nuclear material from peaceful nuclear activities to the manufacture of nuclear weapons or of other nuclear explosive devices or for purposes unknown, and deterrence of such diversion by the risk of early detection.
>
> *Art. 29:* To this end the Agreement should provide for the use of material accountancy as a safeguards measure of fundamental importance, with containment and surveillance as important complementary measures.
>
> *Art. 30:* The Agreement should provide that the technical conclusion of the Agency's verification activities shall be a statement, in respect of each material balance area, of the amount of material unaccounted for over a specific period, giving the limits of accuracy of the amounts stated.

Furthermore, Articles 7 and 32 of the same Agency "Blue Book" make explicit reference to the requirement that the state shall establish and maintain its own system of accounting and that this sytem shall be based on a structure of "Material Balance Areas."

Beside the risks associated with diversion of nuclear materials, a new and serious concern has been developing recently about the possibility of theft of nuclear materials or sabotage of installations. Protective measures against such actions belong to what is known as "physical protection" and are recognized as indispensable and complementary to the set of measures based on quantitative accounting. Physical protection, however, will remain in the hands of the authorities of each state. This is mainly because only the state controls the intervention forces on its territory.

By way of example, Figure 6-1 illustrates the partition of safeguards responsibility in Belgium, and may be compared to the corresponding institutions in the United States. In discussing the respective roles of the two systems of safeguards and of physical protection in the overall coverage mechanism, Farmer's "preconditions" relating to the study of safety (see Chapter 7, below), which apply textually to safeguards, are summarized here. Farmer:

a. Stresses the need for technical understanding in great depth.

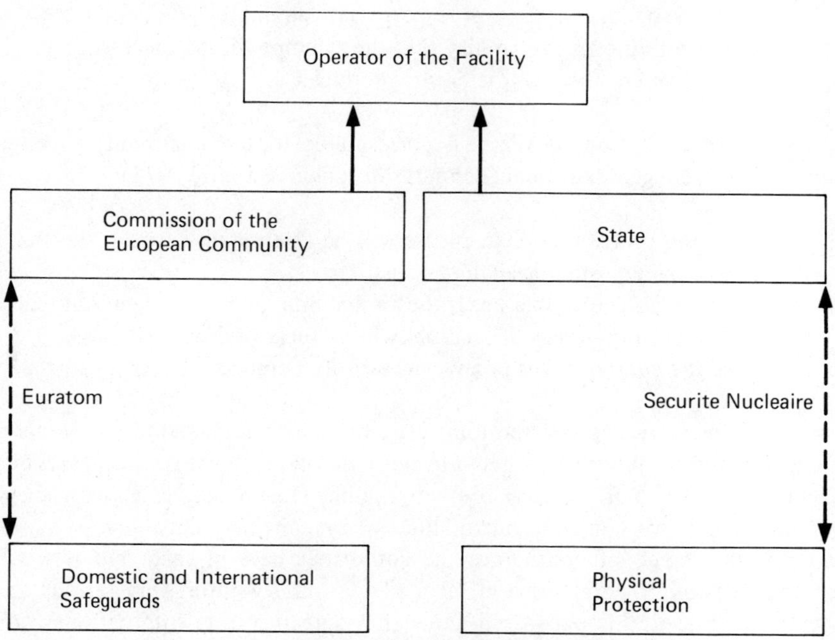

Figure 6-1. Safeguards Responsibilities.

b. Warns of the dangers of premature rulemaking.
c. Calls for an overall view relating nuclear to other risks.

This last parallels the up-to-date lines of thought on a safeguards-system analysis to be found in the basic document by Bennett, Murphey, and Sherr (1975), "Societal Risk Approach to Safeguards Design and Evaluation" (ERDA-7).

In Rasmussen's (1975) study as well as in Farmer's analysis (Chapter 7, below), the basic approach to the quantification of the risk involved the definition of certain initiating "accidents" whose frequency could be estimated, the possible chain of events following these initial events, their conditional probability of occurrence in the presence of given safety measures, and the determination of the consequences of each chain of events. In contrast to these accidental events, the events of safeguards concern are the results of willful actions on the part of an adversary. The analysis proceeds from the terminal events which are the direct cause of a given level of consequences, and not from the initiating action.

The definition of the risk for a given terminal event is:

$$R = \pi \times p \times c$$

where:

π = frequency of attempt to perpetrate an event
p = conditional probability that the attempt will be successful
c = consequences of the events produced

The above definition of the risk corresponds to the commonly accepted functions of a safeguards system (Bennett, Murphey, & Sherr, 1975):

1. To reduce the probability that misuse will be attempted, through the threat of failure or by any other available means.
2. To reduce, to an acceptable level, the probability that any attempt at willful misuse of nuclear materials or facilities will be successful.
3. To reduce the consequences of any successfully perpetrated event.

Clearly, every risk is covered jointly by the two complementary systems of safeguards and of physical protection, with the exception of sabotage, which of course is relevant to physical protection alone (Bahm, Nagele & Sellinschegg, 1975). The relative contributions of the two systems may depend to some extent on the degree of confidence the authorities have in each, but it would not be justifiable to forego one in favor of the other without very serious consideration. Indeed this paper will show that agreements in international safeguards have already developed a valuable tool.

IMPLEMENTATION OF INTERNATIONAL SAFEGUARDS IN THE EUROPEAN COMMUNITY

(CNEN, 1975).
Accountancy is based on dividing the fuel cycle into suitable material balance areas (MBAs), taking physical inventories at practicable intervals, and using suitable accounting reports and records. The first principle of material control is the need for a strict correspondence between the records and the actual physical transactions. The second principle is to obtain a material balance in which all known material movements or inventories are measured (defined as a "closed material balance"). This means that in addition to the major components, all categories of scrap on inventories and all recognized loss streams should be measured. Wherever these principles are applicable, the "material unaccounted for" (MUF) is the primary means of showing any unrecorded loss or gain of material in a specific MBA during a specified time period:

$$\text{MUF} = \text{book inventory} - \text{physical inventory}.$$

Within the framework of these basic principles, it is recognized that safeguards techniques have to rely to the largest possible extent on the existing procedures applied at a facility for material accountancy, flow measurements, and inventory-taking. These procedures are described initially for each MBA in the design information report for submission to the safeguards authority. Thus the safeguards authority has to use operator's data but, at the same time, has to guarantee that this information is correct.

To help define the way safeguards should be implemented in the European Community under the NPT obligations, representative examples of installations were submitted by the Member States during the negotiations of the Subsidiary Arrangements. Routine inspection efforts were estimated for the two safeguards authorities, IAEA and Euratom respectively, in order to meet their objectives in these representative cases.

The IAEA verification plans are based on the following criteria (Rometsch et al., 1975):

a. The probability of detecting a loss of material is chosen to be 0.95 when in fact the true loss is a defined significant quantity.
b. The probability of falsely concluding that a defined significant quantity is lost when no loss actually occurred is chosen to be 0.05.
c. The defined significant quantities are based on the measurement uncertainty that can be reasonably achieved with today's measurement technology.

Therefore, the state of the art regarding nuclear material measurement provides the first basis for defining a significant quantity of nuclear material to be detected, if missing.

The IAEA has compiled performance data on nuclear material determinations both from facility operators and from laboratories in Member States developing advanced methods. From these data, degrees of accuracy have been derived which apply to any facility covering the same type of production (see Table 6-1).

Technical Working Groups recommended the use as "threshold amounts" against which to compare the results of the agency's verification activity; 1 to 8 kg Pu and 1 to 25 kg contained U-235 per year and per MBA. The upper limits of these ranges represent fast critical masses of weapons-grade materials (IAEA, 1971; 1972). The expected uncertainty would exceed these values for industrial facilities with high throughput. The operator cannot personally state that the value of the material unaccounted for in his or her material balance means a real loss or gain of material unless it is at least twice the value of the uncertainty in the material balance. The verifier can only detect the gain or loss of an amount of material larger than those because of the effect of the combination of the operator's and the verifier's measurement uncertainties (IAEA, 1973; Harry et al., 1975).

Implementation in Fuel Fabrication Facilities

(Busca, Love, & Van der Stricht, 1975)

During the last six years, Euratom has progressively implemented a continuously evolving safeguards system in some of the more important PuO_2 and UO_2 fuel fabrication facilities in the European Community. Based on the principle of independent material balances, the system has been developed,

Table 6-1. IAEA Expected Accuracy (Standard Deviation) of a Material Balance Expressed as Percent of Throughput or Inventory (10).

Type of Facility	*Expected Operator's Accuracy*[a]
Uranium isotope enrichment	± 0.2%
Uranium fuel fabrication	± 0.3%
Plutonium fuel fabrication	± 0.5%
Uranium in power reactors	± 0.2%
Reprocessing, uranium line	± 0.8%
Reprocessing, plutonium line	± 1.0%

[a]These accuracies are expressed as one standard deviation and represent the total random and systematic error components of a material balance; its components stem from all types of measurements involved, such as weighing, metering, spectrometer reading, etc., as well as from sampling, equipment calibration, summing up over several batches and the like.

Source: Rometsch et al. (1975).

Table 6-2. PuO_2 and UO_2 Fabrication Plants Inside the European Community

Plant	*Location*	*Capacity, in tonnes (Te) mixed oxide per year*
Alkem	*GERMANY* Wolfgang	40 Te (LWR) 10 Te (FBR)
B.N.	*BELGIUM* Dessel	30 Te (LWR) 4 Te (FBR)
BNFL	*U.K.* Windscale	10 Te (FBR)
CEA	*FRANCE*	5 Te (FBR)

implemented, and improved steadily with the full collaboration of the operators, taking advantage of the already existing real-time accounting systems required for safety purposes. A number of nondestructive methods of measurement had to be developed, which at times proved to be of direct value in the operation of the plants. On the whole, the reinforced inspection scheme produces almost all the information needed for material balances under the NPT, with the required accuracies.

Implementation in Fuel Reprocessing Facilities

(Arenz & van der Stijl, 1975)

Euratom began safeguarding reprocessing plants with the start-up of the Eurochemic facility in Mol (Belgium) in 1966. The Eurochemic Company was created in July 1959, under the auspices of the European Nuclear Energy Agency, by thirteen of its member countries. The founders of Eurochemic had a dual objective: (1) to build and operate, in common, a plant permitting the reprocessing, in reduced quantities, of the very diverse types of fuel coming from European reactors; and (2) to acquire, in common, technical and economic experience to enable European industry to build plants capable of processing the uranium used in power reactors.

Euratom has developed the system of safeguards at Eurochemic in close collaboration with the operator. Thus, a step-by-step procedure of approval has been worked out, in application of the Euratom Treaty (Art. 78), in such a way as to enable the Inspectorate to gain an extensive knowledge of both the process and the facility.

The first step consists in the submission of the so-called "design information" by the operator. This document contains the plans of the facility and the detailed description of the technical flow-sheet, as well as the chemical and physical analytical methods to be used for nuclear material accounting and process control. After any requested changes have been made, the commission issues a partial approval. The second step is a thorough check during and after construction to verify that the construction does in fact comply with the submitted plans. In the third and last step, the correct functioning of the internal system is verified during the cold test of the facility. Calibration and sampling methods are checked during this period. Later on, it was found useful to extend this period to the "warm test" program as well. The same procedure was applied to all the reprocessing facilities in the European Community (see Table 6-3).

Table 6-3. Reprocessing Facilities Inside the European Community

Plant	*Location*	*Fuel*	*Capacity t U/year existing*
	BELGIUM		
Eurochemic	Mol	LWR-HWR-MGR	100
		or	or
		MTR	1.5
	FRANCE		
UP 1 / SAP	Marcoule	MGR + MTR experimental fuel	
		MCR	
UP 2 / HAO + UP2	La Hague	LWR-HWR	800[a]
AT 1	La Hague	FB	0.4
		FB	
	ITALY		
Eurex-1	Saluggia	MTR	0.3
		or	or
		LWR	25
		MGR	
Itrec	Rotondella	Th + U	4
	UNITED KINGDOM		
Dounreay		MTR	0.5
		or	
		FB	10
Windscale		MGR	
		AGR + LWR + HWR	800[b]
	WEST GERMANY		
WAK 1	Karlsruhe	LWR + HWR	50

[a]Start-up expected 1976-77—capacity for LWR fuel.

[b]Capacity for oxide fuel.

The Euratom experience in methods of verification in the reprocessing facilities is summarized in a report by H.J. Arenz and E. van der Stijl (1975). From the very beginning, the verification system has been based on a structure of "Material Balance Areas"—although they were not explicitly referred to by that name—and on the concept of continuous inspection—a mode of inspection intended to maintain continuity of knowledge about the movements and the inventory of the nuclear materials in the plant. Such a continuity of knowledge may or may not require the continuous presence of inspectors or the continuous observation by instrumental methods, depending on the type and size of the plant; but in the specific case of reprocessing plants, continuous presence has been deemed necessary during the campaigns.

In their report, Arenz and van der Stijl quote some materials unaccounted for values for uranium and plutonium as percentages of the throughputs. In the case of plutonium, although MUF values were much in excess of the predictions, acceptable technical explanations were generally found. These explanations, moreover, were often confirmed during intercampaign washouts or by changes in the plutonium isotopic concentrations in the following campaigns.

The actual uncertainties on the material balances for the uranium and plutonium lines appear to be larger, by a factor of 2, than the IAEA expectations (see Table 6-1). It should be noticed, in this respect, that input and output determinations are the main contributors to the overall material balance uncertainty, and that the independent measurement effort should be concentrated on these key measurement points if a higher degree of independence is required.

Parallel with these field verification activities, considerable research and development work has been effected to improve the verification methods, notably through safeguards-integral experiments and through the application of isotopic correlation techniques (Häfele & Nentwich, 1970). "Integral experiment" is a safeguards exercise applied to a production campaign in an existing nuclear facility and extending over a significant period of time. The objectives of such experiments include the closing of the material balance for the actual campaign, the evaluation of measurement errors, the analysis of operating losses, the testing of different instruments and techniques, and the assessment of the necessary safeguards effort.

The Eurochemic reprocessing plant was selected as the field for several integral experiments, designated MOL I, II, III. The MOL III experiment, in particular, was a joint effort of the European Safeguards Research and Development Association (ESARDA) and other institutions (IAEA, USAEC, ACDA, and others) and took place from January 1970 to July 1971 (Kraemer & Beyrich, eds., 1971). The just completed MOL IV experiment principally aims at extending the application of isotopic correlation techniques to fission product correlations and to the results of dissolution sample analysis (Beets, ed., 1973).

"Isotopic correlation techniques" are those techniques that take advantage of the existing relationships between spent-fuel isotopic data to generate an independent estimate of the plutonium content at the input to a chemical reprocessing plant. Figures 6-2, 6-3, and 6-4 are representative examples of existing heavy isotope correlations (Beets, ed., 1973; Christensen, Schneider, & Stewart, 1972; Natsume et al., 1975).

Figure 6-3 concerns a complex ratio:

$$\frac{Pu/U}{(^{239}Pu)\,(^{242}Pu)/(^{240}Pu)^2}$$

This ratio was developed during the analysis of Humboldt Bay fuels and was shown to be insensitive to void variations in a Boiling Water Reactor. In the figure, the Pu/U ratio values are plotted against the $(^{239}Pu)\,(^{242}Pu)/(^{240}Pu)^2$ values.

For the application of these isotopic correlation techniques, the safeguards authority simply has to accumulate the relevant batch data from the diffusion plant, from the fabrication plant, and from the reprocessing plant inputs, and then apply the established relationships to present and future spent fuels of similar nature. The safeguards benefits can even be increased by complementing the empirical data with theoretical calculation to predict the possible variations in the relationships due to changes in the fuel design or in the irradiation parameters. Extensive theoretical investigations are in progress at Battelle Memorial Institute, Richland, Washington; at the Gesellschaft fur Kernforschung, Karlsruhe; and at the Joint Research Center, Ispra (CNEN, 1975: 193).

F.J. Woodman (CNEN, 1975: 221) expressed the idea that, while some of the more firmly established correlations are already in routine use by the reprocessor, a data bank–or pool of information–is necessary to make the potentialities of the technique fully effective, and that, in the interest of efficiency and conservation of resources, there should be only one such data bank in Europe, whether it be operated by Euratom or by the IAEA. Data banks are currently being assembled both at the Joint Research Center in Ispra (CNEN, 1975: 423) and at IAEA headquarters in Vienna (Sanatani & Siwy, 1975).

CONCLUSIONS

The performances of nuclear materials accountancy in safeguards were correctly assessed in the various modern safeguards analyses (Hafele & Nentwich, 1970; Bennett & Granquist, 1967). They will be further improved in the future high-throughput installations, when real-time accountancy and continuous sampling are introduced, thereby making immediate inventory call-out possible at any time. Nevertheless, the role of the complementary surveillance and containment measures will also be enhanced as the throughputs are increased (CNEN, 1975: 405). In particular, the difficult problem of containment at the head-end of reprocessing plants has been investigated by the Karlsruhe group (Hagan, Herre, & Gupta, 1970).

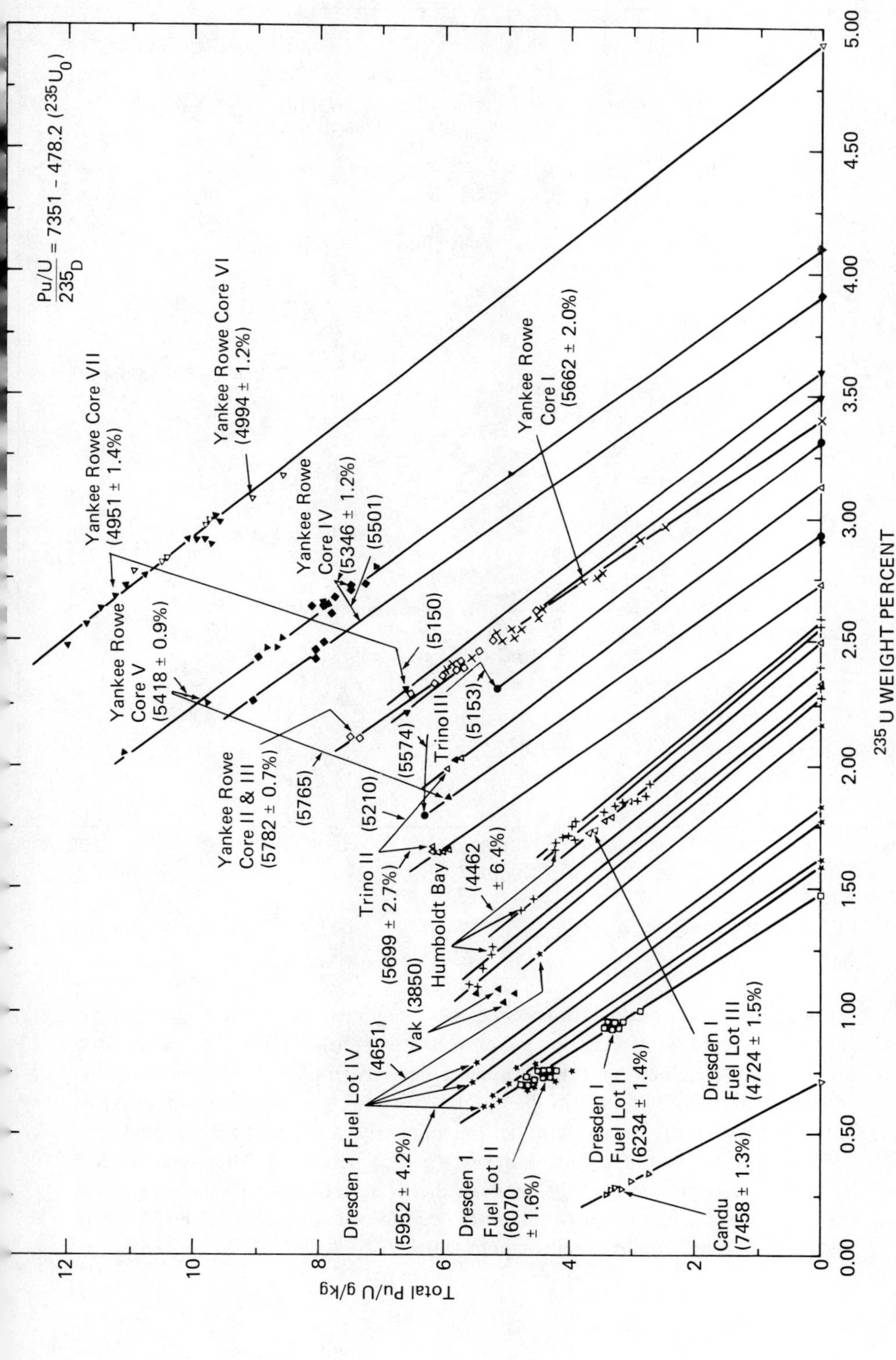

Figure 6-2. Pu/U Ratio Versus ^{235}U Weight Percent for the Collective Data.

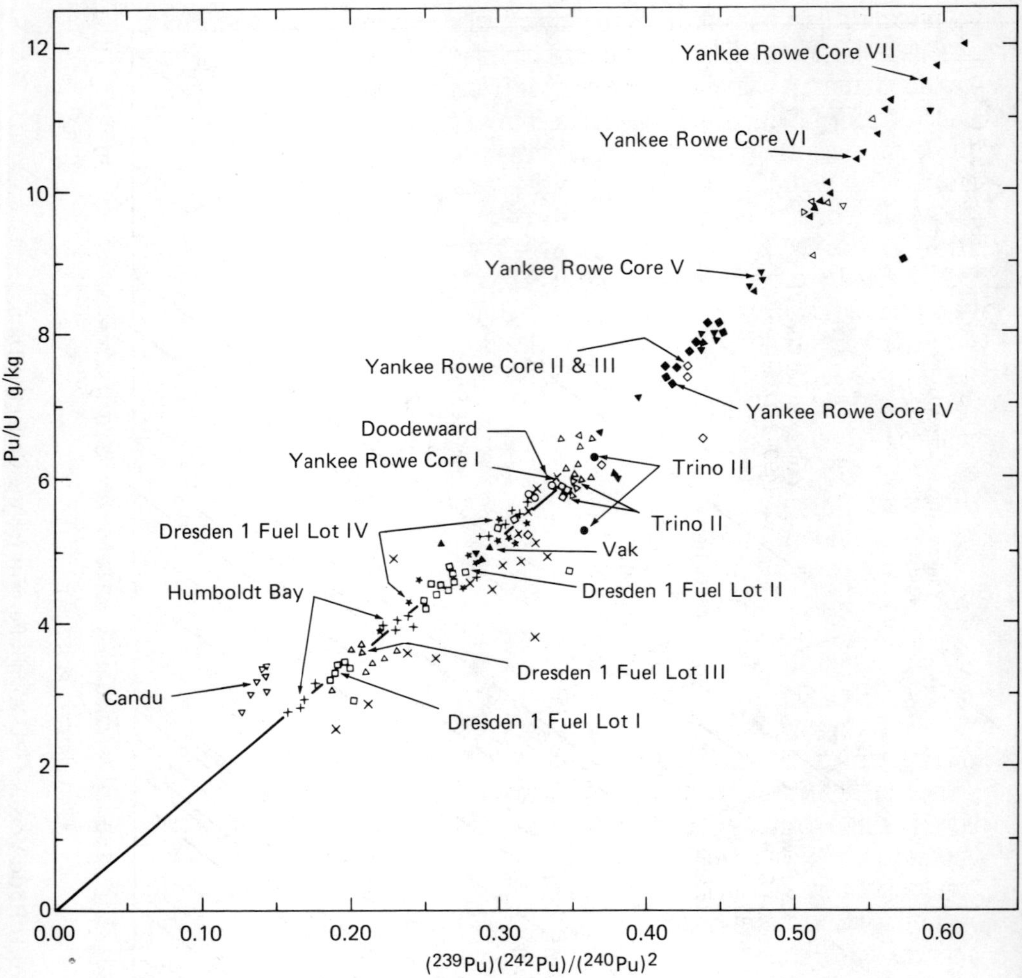

Figure 6-3. Pu/U Ratio Versus $(^{239}Pu)(^{242}Pu)/(^{240}Pu)^2$ for the Collective Data.

The safeguards considerations concerning regional fuel centers are well explained in the paper of P.J. Dyne et al. (Chapter 8, below). It should be stressed that the first constraint on international safeguards arrangements in any regional system is that they will have to be strictly compatible with those of the IAEA. The link between the two systems lies preeminently in material accountancy.

Returning to Farmer's preconditions (p. 95, below), further remarks will be made based on the analogy between the projected regional centers and the structures of the European Community, considered as a regional entity comprising nuclear and non-nuclear weapons states.

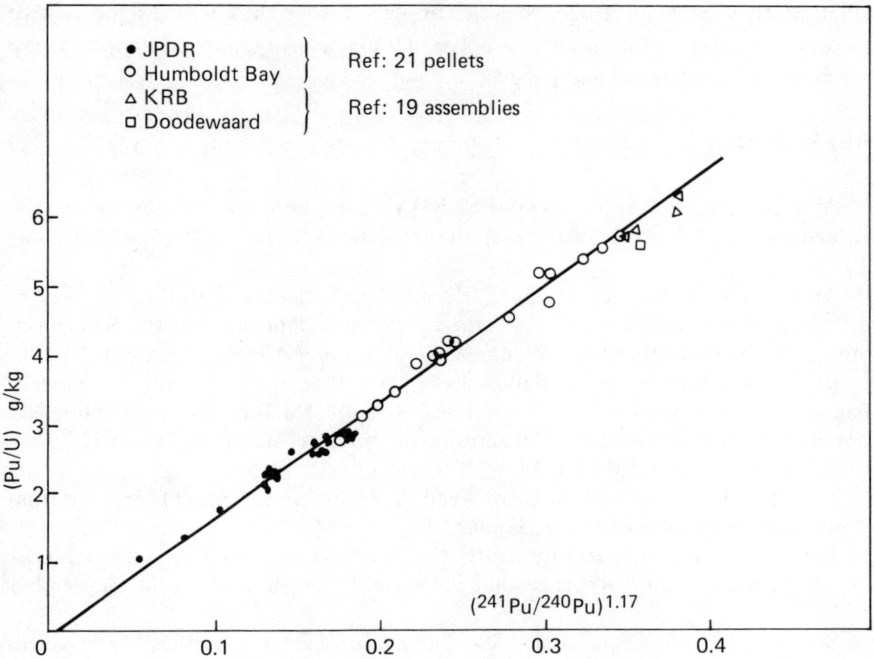

Figure 6-4. Pu/U Ratio Versus $(^{241}Pu/^{240}Pu)^{1.17}$.

The answer to precondition *a*—the highest degree of technical understanding—may be found in the establishment of the European Safeguards Research and Development Association, (ESARDA), which coordinates all the research and development work carried out on safeguards, both in the laboratories of the Commission of the European Communities and of national institutions within the community (Anderson, 1975). It is a unique forum in which representatives of the safeguarding authority (Euratom) and of the research and development laboratories can keep themselves mutually informed, at regular intervals, of their latest developments and projects. Furthermore, because the research and development groups are also in touch with the nuclear plant operators in their countries, an effective two-way channel of communication exists between the latter and the central authority.

As to precondition *b*—the dangers of formulating premature laws and rules—the IAEA and the international community would probably be wise to watch the operation of the new Euratom regulations established in compliance with the Rome Treaty and with the NPT.

Concerning precondition *c*—an overall approach in safety problems—the ERDA-7 report (Bennett, Murphy & Sherr, 1975), apparently provides the appropriate framework. It points out that the safeguards system has, in general, many stages at which it can foil the action sequence of the presumed adversaries, whether they attempt force, stealth, or deceit, and thus reduces the risk to society. Moreover, the report supplies a long list of appropriate protective mechanisms, both active and passive.

REFERENCES

Anderson, A.R. "Activities of ESARDA" International Symposium on the Safeguarding of Nuclear Materials, Vienna, 20-24 October 1975. IAEA–SM/201/82.

Arenz, H.J. & van der Stijl, E. "Euratom Experience of Methods of Verification in Reprocessing Facilities". International Symposium on the Safeguarding of the Nuclear Materials, Vienna, 20-24 October 1975.–IAEA/SM/201/70.

Bahm, W., Nagele, G. & Sellinschegg, D. "Risk Classification for Nuclear Facilities in Connection with the Illegal Use of Nuclear Materials". International Symposium on the Safeguarding of Nuclear Materials, Vienna, 20–24 October 1975, IAEA/SM/201/49.

Barnaby, F. Preventing Nuclear Weapon Proliferation. Stockholm: International Peace Research Institute, January 1975.

Beets, C. editor Contributions to the Joint Safeguards Experiment MOL–IV at the Eurochemic Reprocessing Plant, Mol, Belgium BLG–486 September 1973.

Bennett, C.A. & Granquist, D.P. "Safeguards Systems Studies" in Proceedings of the Symposium on Nuclear Materials Safeguards, Argonne National Laboratory, June 26–27, 1967. WASH–1076.

Bennett, C.A., Murphey, W.M., and Sherr, T.H. Societal Risk Approach to Safeguards Design and Evaluation. ERDA-7, 1975.

Busca, G., Love, B. & Van der Stricht, E. "Euratom Experience of Methods of Verification in Fuel Fabrication Facilities". International Symposium on the Safeguarding of Nuclear Materials, Vienna, 20–24 October 1975.–IAEA/SM/201/71.

Christensen, D.E., Schneider, R.A., & Stewart, K.B. Summary of Experience with the Use of Isotopic Correlation Safeguard Techniques. BNWL-SA-4273 (March 20, 1972).

CNEN. Practical Applications of R & D in the Field of Safeguards. Proceedings of a Symposium sponsored by the European Safeguards Research and Development Association, Rome, March 7-8, 1974. Edito del CNEN, Roma, 1975.

Farmer, F.R., "Safety Considerations of Nuclear-Fuel Processing Plants" Chapter 7, below.

Häfele, W. & Nentwich, D. Modern Safeguards at Reprocessing Plants and Reactors: Safeguard Techniques. Proceedings of a Symposium, Karlsruhe, 6–10 July 1970. IAEA/SM/133/103 (STI/PUB/260.

Hagen, A., Herre, F. & Gupta, D. Design Criteria for Reprocessing Plants KFK 1107 (November, 1970).

Harry, R.J.S., Anderson, A.R., Rota, A., Kraemer, R. Kschwendt, H. Practical Aspects of Nuclear Material Accountancy and their Applications to Safeguards. In CNEN (1975) above.

IAEA, 1971. Report of the Working Group on Quantitative Data and Results of Systems Analyses and Integral Tests. Vienna 4–8 October 1971. IAEA-PL-455.

IAEA, 1971 a. Structure and Content of Agreements Between the Agency and States Required in Connection with the Treaty on the Non-Proliferation of Nuclear Weapons. INFCIRC/153, May 1971.

IAEA, 1972 Report of the Working Group on Accuracy of Nuclear Material Accountancy and Technical Effectiveness of Safeguards. Vienna 28 August –September 1972. IAEA-PL-488.

IAEA, 1973 Report of the Panel on Systems of Accounting for and Control of Nuclear Materials. Tokyo 5–9 November 1973. IAWA-PL-581.

Kraemer, R. & Beyrich, W. editors. Joint Integral Safeguards Experiment (JEX 70) at the Eurochemic Reprocessing Plant, Mol, Belgium (January 1970 to July 1971). EUR-4576, or KFK-1100) August 1971.

Natsume, H., Matsuura, S., Okashita, H., Umezawa, H. & Ezure, H. "Gamma-Ray Spectrometry and Chemical Analysis on JPDR-I Spent Fuels." JAERI-memo-6291 (October 1975).

Rasmussen, N.A. Reactor Safety Study—An Assessment of Accident Risks in U.S. Commercial Nuclear Power Plants USAEC, October 1975 (WASH-1400).

Ribicoff, A. & Percy, C. Peaceful Nuclear Exports and Weapons Proliferation: A Compendium. Washington; Government Printing Office 1975.

Rometsch, R., Lopez-Menchero, E., Ryzhov, M., Hough, C.G. & Panitkov, Y. "Safeguards—1975–1985." International Symposium on the Safeguarding of Nuclear Materials, Vienna, 20–24 October 1975. IAEA/SM/201/103.

Sanatani, S. & Siwy, P. "IAEA Bank of Correlated Isotopic Composition Data." International Symposium on the Safeguarding of Nuclear Materials, Vienna, 20–24 October 1975 IAEA/SM/201/100.

USACDA. The Danger of Nuclear Proliferation. U.S. Arms Control and Disarmament Agency Publication 1975, released November 1974.

Willrich, M. & Taylor, Th. B. Nuclear Theft: Risks and Safeguards. Ballinger Publishing Company, Cambridge, Mass.: 1974.

Chapter 7

Safety Considerations of Nuclear-Fuel Reprocessing Plants

F.R. Farmer

PRECONDITIONS FOR INTERNATIONAL SAFETY PROVISIONS

If there is to be useful international oversight of nuclear safety, certain preconditions would be essential:

a. The problems, the nature of the risks, and means for their modification should be well understood. This requires the highest degree of technical understanding at great depth in many disciplines–physics, chemistry, engineering–and in practice as well as theory.
b. There should be clear recognition of the dangers of formulating laws, rules, and practices at too early a date. Several countries, including major ones, have endeavored to ensure nuclear safety by rule; the setting up of a control body usually leads to rules–to justify the existence of the body and to give superficial appearance of control to satisfy the politician.
c. Whatever problems are discussed in nuclear safety should be seen in relation to the risks in modern society–not in isolation.

As an amplification of these points, I prepared a paper for a recent meeting of Accademia Nazionale dei Lincei, Rome, which will be drawn upon here. A recent report of Sir Edward Pochin, "Estimated Population Exposure from Nuclear Power Production and other Radiation Sources," published by the Nuclear Energy Agency of the OECD in January 1976 will also be cited.

This paper is addressed to the possible consequences of accidents in a reprocessing plant, compares these with the "safety" reports for reactor installations, and comments on some aspects of risk assessment and licensing in a multinational situation. In developing this theme, I will endeavor to show the difficulty of describing safety by referring to reactors where we have a history of twenty years of discussion, meetings, reports, and so on, much of which has a bearing on reprocessing plants.

MAXIMUM CREDIBLE ACCIDENT

It is not easy to give a meaningful description of the hazard potential of nuclear installations (or equally for some non-nuclear installations). There is not a unique accident, and for any one accident there is not a unique consequence. Now, the nuclear community—particularly in licensing parlance—tried to overcome this difficulty by defining a maximum credible accident and sanctified this concept in guide rules.

Measures were then taken to ensure that the consequence of a maximum credible accident in terms of hurt to people would be only trivial—as may be derived from WASH-1250 (USAEC, 1973)—and in line with the doctrine that a reactor may be licensed if it can be shown that the installation will cause no hurt to people.

Another picture emerges from the consideration of accidents of greater severity than the maximum credible, as in papers at the first Geneva Conference (McCullough et al., 1955; Parker & Healy, 1955; Marley & Fry, 1955), WASH-740 (USAEC, 1957), and recently in WASH-1400 (Rasmussen, N., 1975).

We find from WASH-740 that an accident might kill 3000 people; from the first Geneva Conference, Parker estimates 200–500 for a particular population density, and Rasmussen gives a wide range in keeping with a more probabilistic approach. The question is often posed, "What is the worst that can happen?"—not inconsistent with the logical interpretation of the maximum credible accident. As the answer ranges from near zero to thousands or tens of thousands of casualties, it is no wonder that nuclear installations are thought to have a danger potential far greater than other industrial installations, but that idea has been implanted by the question which is not posed about more ordinary explosive and chemical plants, even some that have indeed triggered major disasters.

ASSESSMENT OF HAZARD AND EFFECTS OF ACCIDENTS

In discussing hazard we must provide a framework that can be understood by the serious reader. To illustrate my point I choose an example from a non-nuclear hazard—perhaps more easily appreciated and less emotive. Westbrook (1974) made a careful study of accidents to road and rail tanker traffic to assess the likely consequences of carrying chlorine by tanker. Using best data on accident frequency, severity, average route, population density, time for rescue teams to arrive, average weather—he concludes that the chlorine traffic in the United Kingdom might kill one person or so in fifty years. I agree with most of this analysis and broadly with the conclusion—as a "*likely*" result. In the nuclear field we ask, "What if?," and these questions pose a series of more improbable happenings. The counterpart in the chlorine story would be to describe the consequence of releasing perhaps 50 percent of the tanker in a short period of time, in a populated area, and with little benefit from evacuation. I then estimate a 10 to 50 percent probability of 200-500 deaths. I do not expect this to happen, but it could.

The attitudes to nuclear plant safety have thus been conditioned not only by the bomb, but by the description of unlikely events, or combinations of event, weather, population, and so on. Not so much that these are technically wrong, but rather that we are not used to considering the whole range of possible accidents and their consequences and concentrating on the more extreme ends of these distributions. This is not generally undertaken for non-nuclear plants, but if carried out would yield numbers of potential casualties of the same order as, or greater than, nuclear.

Let me summarize the consequences of a serious reactor accident and then move to reprocessing plants.

Most thermal reactors could have accidents which lead to the melting of the reactor fuel. This makes available the more volatile fission products, Xenon, Krypton, Iodine, Tellurium, and some ruthenium. I believe the worst release to atmosphere arises from 10 percent of the total volatile fission products. The consequences on a populated U.K. site, some 5 to 10 times more populated than Rasmussen's average site, might be described as:

50 percent chance of zero casualties.

30 percent chance of 100-1000 cases of thyroid disorder.

20 percent chance of 1000-10,000 cases of thyroid disorder these occurring in 10-20 years after the accident, and carrying some small risk of death; additionally, from 10-200 deaths from other radiations mostly delayed for 10-20 years.

contamination of ground, requiring control of occupation and farming—to a serious extent up to 10 miles, to a lesser degree perhaps 30 miles.

SPENT-FUEL REPROCESSING

A slightly different picture emerges from chemical reprocessing—radioactive-iodine decays—for most of the process, the longer-lived fission products and plutonium and actinides are in solution. These solutions are processed in totally enclosed or confined plant areas with high-grade gas-filters; appropriate sections of the plant are retained within lined and shielded structures designed to prevent escape of liquid or gas. The total installation will provide for the storage of the incoming fuel, for intermediate and terminal products; material will be processed in passing from one storage area to another. In general, material in store is in a well-defined situation and can be extremely well monitored and protected. It is just possible that some gas, volatile fission products, or even liquid could escape through progressive failure in protection, but such escapes are likely to be local in effect, and may be costly and cause the plant to be shut down. From these considerations, storage areas need very careful appraisal by a competent management.

The liquid solutions in process will be passing through a series of columns, tanks, pipes, and filters,* and it is difficult to foresee, even from a complete pipe break, any gross release to atmosphere of liquid or vapor.

A hazard could arise at a distance of many miles from the plant if the radioactive material were to be transported by air or by water. If by water, people can be alerted—the result may be costly but is unlikely to cause many casualties. If by air, it is much more difficult to move people quickly, and inhalation is by far the greatest danger.

How can materials be dispersed in air from reprocessing plants? Conceivably by boiling; this may cause a preferential release of volatile fission products (as in reactor accidents), but this mechanism does not seem to me to be the basis of a major catastrophe.

A large amount of energy is required to disperse as vapor, or fine mist, a large volume of liquid; this could conceivably come about through chemical interactions but, even so, an escape from the process should be largely contained within the surrounding structures.

An assessment of these possibilities would need to be carried out for a specific design. In more general terms—as in reactor assessment—we can ask, "What is the worst that might happen?"—"Worst" being conditioned by assumptions about quantities of material in process and by site condition.

The quantity of material in process is relatively small—I cannot foresee the release of a greater volume of liquid than that equivalent to one tonne of fuel—and even this is probably a gross overestimate. Hence, reprocessing plants may have a higher probability of causing a costly local contamination than reactors, and perhaps a higher probability of generating small and locally dangerous releases of gas or vapor, but a low probability of causing a large dispersed atmospheric release. I assess that the quantity which may be dispersed is smaller than from a serious reactor accident but different in composition, comprising ruthenium and other fission products, plutonium, and actinides. The potential harm, curie-for-curie, is greater than for a reactor's volatile fission products, and is roughly equal as between fission products and actinides.

*See Figures 3-3 and 3-4, in Chapter 3, above.—Eds.

ASSESSMENT OF PLUTONIUM

Plutonium does not dominate the risk scene as popularly supposed. On many occasions I have read or heard the comment that plutonium is the most toxic material. Even recently Cochran (1974) has described plutonium as "one of the deadliest substances known to man." He continues,

> As noted by the International Commission on Radiological Protection (ICRP), in terms of amount available, projected usage, extent of anticipated accidental human exposure, and radiotoxicity, plutonium is the most formidable radionuclide in the periodic table.

Cochran's interpretation of the ICRP statement is not true. I do not wish to underestimate the potential hazard of plutonium or the concern that people may have about the development of a so-called plutonium economy, but it is important in reviewing these that we do not rely on emotional statements or hearsay but use the best scientific information available to us. We may well be swayed by the type of remark made in 1955 by McCullough (1955):

> Radioactive poisons are more hazardous than chemical poisons by a factor of 10^6 to 10^9.

This is not true for plutonium. If we compare the weight of plutonium or chlorine required to cause death (quickly for Cl, say one year for Pu), then the ratio is about 1 to 10; 1 mg Pu or 10 mg Cl. If we consider the dose which might—on a linear hypothesis—give a 0.1 percent chance of death after many years, then we arrive at a lower figure for the quantity of plutonium by a factor of 1000, but no figure could be set for chlorine. We should not forget that there is a risk of fatal accident in a wide range of industry of about 3 per 1000 over the period of employment.

Returning to the reprocessing plant accident, the resulting harm depends on the site—fewer reprocessing sites are needed than reactor sites and they may be in rather less populated areas. If I take the site characteristic of Rasmussen's average of 66 sites, I would estimate the effects of the release described above to lead to 50-200 deaths and ground contamination due to cesium and actinides requiring rather prolonged control over 10-15 miles. So reprocessing plants—in common with reactors and other industrial processes—can constitute a serious hazard, but one which can be reasonably assessed both qualitatively and quantitatively. It is a process in which the risk of harm should be low, provided the design and operation are well managed.

SAFETY ASSESSMENT AND LICENSING

There is a fair measure of agreement about the standards required for normal operation of nuclear plant; that is, as to radiation levels, means for protecting workers, means for measuring discharges, and so on. These standards are still evolving but I do not believe that any small difference in procedures from country to country would cause difficulty in a multinational scheme. If the benefit is worthwhile, small differences should be resolved. However, there is no blueprint for safety in avoiding possible accidents in nuclear plants. Each plant must be assessed, and in reprocessing plants we cannot proceed on the basis of past decisions, as we can for reactors.

The assessment of a design and proposed operation and maintenance procedures calls for a degree of expertise not readily available. If the plant were to be built in a country with considerable experience with nuclear power, then it would be subject to licence within the country and assessment by its own experts.

FIRST PLANT IN A HOST COUNTRY

If the plant were to be built in a country with no experience of reprocessing, a new situation arises:

1. It would need to be approved and licensed by the host country.
2. The host country would not initially have the range of experience and expertise to assess the proposal.
3. The host country could assemble a small team of people who could receive, criticize, and accept—or in part modify—an assessment carried out by an external team of trained people.
4. It is unlikely that an existing licensing group would provide this external team, having their own internal duties and allegiances.
5. The type of visiting team brought together by the IAEA for past siting exercises or reactor reviews would not be suitable for the task.
6. a. The team required should comprise some of the most competent people in assessment work—engineers, physicists, and chemists—of varying experience, but *not* those who can best be spared.

   b. The team would need a hard core wholly involved for 1-2 years—say six to eight people and another ten people providing effort to a program amounting in all to 30 man-years at a cost of perhaps $1 million.
7. Such a team could be assembled and managed through the IAEA if there were sufficient concern among countries to make the project a success.
8. The team initially assesses the broad design objectives and later considers the detail in conjunction with the major design contractor, recognizing that most of the safety and reliability assessment must be carried out by

the designers. The assessment team does not detract from the design effort or responsibility—it should supplement and aid the designer.

Later the team assesses the means for quality control in manufacture and installation, and progressively enters into discussion with the operator, who is responsible for commissioning and operation procedures, including maintenance manuals.

Finally, schemes for general supervision, periodic reviews and information feedback are to be agreed upon with the operator.

REFERENCES

Cochran, T.B. *The Liquid Metal Fast Breeder Reactor—An Environmental and Economic Critique.* Washington, D.C.: Resources for the Future, Inc., 1974.

Marley, W.G., and Fry, T.M. "Radiological Hazards from an Escape of Fission Products and the Implications in Power Reactor Location," Proceedings of International Conference on the Peaceful Uses of Atomic Energy, Geneva, 1955.

Parker, H.M., and Healy, J.W. "Environmental Effects of a Major Reactor Disaster," Proceedings of International Conference on the Peaceful Uses of Atomic Energy, Geneva, 1955.

McCullough, C. Rogers, et al. "The Safety of Nuclear Reactors," Proceedings of International Conference on the Peaceful Uses of Atomic Energy, Geneva, 1955.

USAEC, 1957 *Theoretical Possibilities and Consequences of Major Accidents in Large Nuclear Power Plants.* (WASH-740).

USAEC, 1973 *Safety of Nuclear Power Reactors (Light-Water Cooled) and Related Facilities.* (WASH-1250).

USAEC, 1975. Rasmussen, N. *Reactor Safety Study—An Assessment of Accident Risks in U.S. Commercial Nuclear Power Plants.* (WASH-1400).

Westbrook, G.W. *The Bulk Distribution of Toxic Substances: A Safety Assessment of the Carriage of Liquid Chlorine.* Proceedings of Symposium "Loss Prevention and Safety Promotion in the Process Industries." Elsevier Scientific Publishing Company, 1974.

Chapter 8

Regional Fuel-Cycle Centers: Technical and Operational Considerations

P.J. Dyne
R.B. Lyon
D.R. McLean

A convenient starting point for discussion of regional fuel-cycle centers is to examine the considerations which go into a decision to establish a national fuel-cycle program. We will use, as a basis, a number of studies we have been making on the timing and logistics of establishing a reprocessing industry in Canada.

Over the next few decades we expect Canada to be building more heavy-water-moderated reactors, fueled by natural uranium. The time at which reprocessing to recycle plutonium becomes economically attractive will depend on the price and availability of uranium and the demand for expansion of nuclear generating capacity. For heavy-water reactors, recycle of plutonium is one step toward the development of the thorium uranium fuel cycle, which offers a large extension of the resource base. The time at which this is economically attractive again depends on the price of uranium and the need for nuclear power. In the interim the spent zirconium-clad fuel can be kept in retrievable surface storage for at least fifty years.

The decision in Canada to implement either plutonium recycle or the thorium uranium cycle and a commitment by a utility to operate reactors using plutonium or uranium-233 as fuel could only be taken after demonstration, on a significant scale, that reprocessing, fuel fabrication, and fuel and reactor performance were economical and reliable. Since security of supply is so important to a utility, reliability will be heavily weighted. If the utility were large enough, they might insist that they own and operate the fuel-cycle plant.

If the recycle of plutonium were the only objective, a fuel reprocessing plant could be purchased from one of a number of organizations who should be able to give guarantees of performance and reliability. However, in order to be an intelligent customer for the plant as well as a competent operator, it would have to establish a significant domestic research and development (R&D) effort to form the nucleus of the operating and supporting laboratories. Our experience with other nominally developed technologies, heavy-water production being an example, is that they cannot be efficiently operated in another country without the host country having detailed technical expertise in that technology. In addition there would be a domestic R&D program for the fabrication and testing of mixed oxide fuel for CANDU reactors and a demonstration operation of a CANDU using such fuel.

For other reactor types the decision to establish a reprocessing industry may depend on other considerations. Fuel from the gas-cooled MAGNOX reactors, developed in the U.K., cannot be stored indefinitely because the magnox alloy corrodes slowly in water. Reprocessing is thus a requirement for waste management but also provides a supply of plutonium for fast-breeder reactors. Spent fuel from light-water reactors (LWR's) could be stored indefinitely. The incentive for reprocessing the fuel from these reactors is a combination of the need for plutonium for recycle or for use in fast reactors and the value of the enriched uranium in the spent fuel.

A large fraction of the cost of a reprocessing and fuel-fabrication facility is associated with waste management activities, control of effluents, solidification of high-level (fission product) wastes, their final storage or disposal, solidification and storage of α-active wastes from fuel-fabrication plants. It is essential that, at the industrial stage, every one of these activities is taken to completion, no material being allowed to accumulate that will have to be treated or handled or disposed of in some unspecified manner "at a later date." While interim storage of fuel, pending a decision to reprocess, is logical and politically acceptable, interim storage of wastes pending undefined decisions on their final fate is not, in our opinion, sensible or politically acceptable. We believe that this consideration applies with equal force to the proposed regional fuel-cycle centers.

The siting of the plant and the associated waste management and fuel-fabrication facilities will require careful study. Many will argue that for safety and security reasons the plant should receive spent fuel and produce new fuel within the boundaries of its own plant fence; separated fissile material should not be moved in the public domain from a reprocessing plant to a fuel-fabrication plant. This concept might, with advantage, be taken even further. Any fissile material, plutonium or U-233, should be mixed in situ with U-238 or thorium so that at no point in the process is there an accumulation of separated fissile material. For reasons of economics it would also be desirable to have the waste disposal site at or close by the reprocessing plant. As we, in Canada, do not expect such facilities to be in operation before the turn of the century, by which time a considerable inventory of spent fuel will have accumulated, we favor the concept of central fuel-storage facilities at the same site as the fuel-cycle plant.

THE REGIONAL FUEL-CYCLE CENTERS

Incentives

In this section we review the incentives that might exist for a number of countries to organize a regional center, and review some of the many technical, organizational, and political matters that would have to be resolved.

While this paper deals with technical and operational considerations, it is idle to pretend that the primary inducement for the construction of regional cycle centers will be other than political. International peace-of-mind over weapons proliferation is the primary objective. One would hope that participating countries will wish to be seen to be involved in a safeguarded and internationally monitored use of fissile material. Further, one can anticipate in the interests of nonproliferation that major powers already possessing nuclear weapons may exert political pressures which would make participation in regional centers to be seen to be in the national interests of the participating countries.

These political factors must, however, be exerted in a framework of economic and industrial practicality. The electric utilities in each nation must have a requirement for plutonium recycle, other advanced fuel cycles, or the need to recover enriched uranium from LWR fuel. This requirement would depend on an assessment of their respective load and load growth and the projected price and availability of natural and enriched uranium. In principle the governments have a choice between establishing a domestic fuel-cycle capability or of joining with other nations in a consortium for a regional center. With economies of scale and the sharing of costs on overheads and support facilities, which might be insensitive to the size of the plant, reprocessing and fuel-fabrication costs should be lower for large regional plants than for small national plants.

General Plant Considerations

Assuming that these political and economic questions were resolved so that a consortium was established, the participating nations would have to set up a joint design and engineering team to specify the plant. We will assume that the consortium would then approach one or more engineering firms for proposals for its design and construction. The specifications would include:

1. *Type of fuel to be processed.* Probably fuel from more than one sort of reactor with differing plutonium and uranium content and with differing burnups will have to be accepted, and the fissile material would then have to be refabricated into elements with fissile content appropriate to the various reactor types. For UO_2 fuel this does not present particular problems except in the recovery of uranium from CANDU fuel and from LWR fuel. The U-235 content of spent CANDU fuel is much lower than that of spent LWR fuel and has little value. Uranium from spent LWR fuel has, however,

significant U-235 content and should not be mixed with uranium from spent CANDU fuel. Processing of thorium and fast-reactor fuels could present more of a problem because industrial plants for these operations have not yet been built.

2. *Specification of emission and waste management standards.* For plants which will be built in the future, a large fraction of the cost is associated with waste management, equipment for the solidification of wastes, and systems which control the emissions of radioactive gases as well as the amounts of low-level activity in aqueous streams which can be released to the environment. While there is an international consensus that these releases shall be "as low as economically possible," how low is a matter of debate! However, at the outset of the design study, standards and specifications will have to be set. Whose standards are to be followed? Some countries may have already promulgated standards and design criteria, while others may not have established the legal framework for setting and administering these standards. Whatever standards are set, however, the host country will certainly have the final word.
3. *Safety standards.* Safety requirements also vary from one nation to another or may not have been promulgated. The required standards will also vary widely with geographic location; that is, earthquake or hurricane conditions. However, whatever international safeguards on the operation are agreed upon, the host country must have the final say on matters related to safety.

Disposal of Radioactive Wastes

As noted in the earlier section, we regard it as mandatory for any future plants that a complete plan for managing and disposing of radioactive wastes be established at the time the plant is commissioned. Fuel-reprocessing plants and fuel-fabrication plants as now envisaged will have, in general, two major waste streams. First, concentrated fission products, solidified into a glass matrix; and second, residues of long-lived α-emitting heavy elements (plutonium, americium, and curium, and the like), which are left in the reprocessing streams or which appear in the fuel fabrication plants. These would also be converted to a solid form.

The level of radioactivity of the concentrated fission products is such that the wastes generate significant amounts of heat; these wastes cannot be stored or disposed of without provision for cooling. These cooling requirements diminish with time and are negligible after about two hundred years. On the other hand, the α-emitting wastes which do not generate much heat remain significantly active for many thousands of years.

High-level fission-product wastes could be stored in engineered structures at or near the surface. Such engineered structures could always be monitored and the storage could be done in a retrievable fashion so that the material could be later moved to a permanent disposal site (when one was identified) or to other structures if the present ones deteriorated. For fission products, monitoring and supervision would be necessary for about 500 years. However, there is an international consensus that disposal methods having a minimum of monitoring and supervision are needed for long-lived α-emitting residues and that such disposal methods would also be preferable for the high-level fission-product wastes. To date, most schemes propose burial of the material in some specially chosen geologic formation, bedded salt being the one most frequently discussed. Other concepts involve burial in hard rock, shale, and beneath the bed of the sea. The Federal Republic of Germany is the only country now utilizing geologic disposal, using an abandoned salt mine near the town of Brunswick.

The waste management plans for the fuel-cycle center would have an effect on the acceptability of the project itself, the choice of the host country, and on the siting of the center.

Let us suppose, for the sake of argument, that one country had within its boundaries geologic formations suitable for the disposal of wastes, and that sufficient geophysical work had been done to enable that country's regulatory agency to license a repository in these formations. Since, by hypothesis, the hazard to the population due to the disposal of these wastes was minimal, there would be no well-founded reason why this repository should not take wastes from any other country. The concern, "Why should we look after other people's wastes?" is only valid if there is a significant cost or hazard associated with the acceptance of wastes into a repository. A country which had such a potential repository and such a perception of the risks would, other matters being equal, be the rational choice for the host country.

However, at present the practical realities differ significantly from this situation. While, as indicated above, there is an international consensus that such geologic sites will be developed, their location and development is only likely to be accomplished within the next ten or twenty years. It would thus be very difficult to choose a site *today* for a regional fuel-cycle center with any positive assurance that a suitable disposal site would exist within the borders of the host country.

It is thus likely that, for a regional fuel cycle center planned in the near future, wastes would have to be kept in some interim storage facilities pending a decision on disposal. These wastes would be in a solid form, almost certainly in the glassy matrix proposed by British Nuclear Fuels Limited (BNFL) and could be either stored on the site or sent back to the country of origin of the spent fuel for interim storage. This latter option, while feasible and politically acceptable, seems unnecessarily awkward. In the technical nuclear community there is a strong conviction that such interim storage can be done at entirely adequate levels of safety. There may, however, be a political reluctance to accept such a solution, even on an interim basis, particularly for the host country if it is to be the custodian for the interim period.

Transportation of spent fuel and of radioactive wastes have similar dimensions. Wherever and however fuel-cycle centers are built, spent fuel will have to be brought to them and radioactive wastes, refabricated fuel, or separated fissile material may have to be moved away. The economic penalty for transportation over extended distances is not a large component of total fuel-cycle costs. Nevertheless, for reasons of logistics alone one would argue that transportation of highly radioactive materials should be minimized. For this reason again it would be desirable for the host country to be the site for disposal.

It is beyond the scope of this paper to address the general question of public acceptability of storing or disposing of nuclear wastes. Whatever the technical solutions may be, the acceptability and viability of regional fuel centers, for the next few years, may turn on the political and social aspects of this waste management. This is probably the single most important point in this paper.

Specifications of the Plant to Meet Safeguard Standards

One of the principal motivations for regional reprocessing centers in the international community is that the international supervision implicit in the scheme will provide the necessary guarantees and assurances that fissile material is not being directed into nonpeaceful uses. In effect this can only be done by strict control of the inventory of fissile material entering and leaving the plant and by a study of diversion paths. This control of inventory is, in any case, essential for the efficient operation of the plant. We envisage the member nations being credited with the fissile material contained in the fuel they send for reprocessing and then being charged for the fissile content of the fabricated fresh fuel they purchase from the fuel-cycle facility. It seems unnecessary to restrict the operation in such a way that nations could use plutonium only in amounts they produce, and further that they use only the plutonium they have themselves made (as if the atoms carried their own national flags!). Normal economic requirements and good chemical engineering practice would require a very high standard of operation which would be compatible with the requirements of international safeguards. The international aspect of these controls provides assurance to other countries that these standards are being rigorously met.

From the outset, the capability for safeguards inspection must be incorporated into the plant during the preliminary design. These capabilities include accountancy, containment and surveillance (not unlike the requirements of a bank!). The determination of material balances is an example of the capability for accountancy.

As is well known, however, overall material balances in such plants are impossible to achieve to the accuracy required to give an exact assurance that fissile material is not being diverted. One reason for this is that the balance is derived from a set of concentrations and volumes which vary over wide ranges. To detect diversion it would be prudent to establish material balances in several portions of the plant, particularly those places where plutonium is being handled in significant concentrations. These methods of assay should be adopted on an international basis together with international agreement on the precision required for these mass balances. In general, similar questions arise over the security of the plant. This would again have to be arranged on an "international" basis with, probably, a major input from the host nation. There is a need for R&D in developing methods for the continuous assay of plutonium, for example, and in all other safeguarding techniques.

The Choice of the Host Country

We have already touched on a number of considerations that would influence a group of nations to form a consortium. These include the size of their domestic installations, the compatability of the reactor types and fuel designs, their geographic proximity, their political and cultural alignments and similarity or commonality of languages, and their mutual interest in nuclear safeguards.

The choice of the host country among members of the consortium turns upon a number of other considerations which often on examination prove to be two-sided. The existence of significant industrial and nuclear capability is an excellent example. A nation which already has relevant industrial capability would, in a normal commercial situation, be in a very strong position to bid for the siting of the plant, since it can give good assurances that the plant can be built to schedule and cost, can be operated economically and reliably, and the necessary support functions can be performed rapidly and efficiently. The higher wages in industrialized countries may be more than offset by higher productivity.

International interest in industrialization and technological development might, however, argue in the contrary fashion. To counter the trend of the rich getting richer, the least industrially and technically developed country should be chosen as the site. The host country could benefit from the large capital investment being made in its territory and the significant increase in local employment required for the operation of the plant. This is, however, not the proper place to argue the pros and cons of industrial development in less-developed countries.

The matter of possible political control of the plant by the host country presents several delicate political problems. While it might appear to the host country that it is in an advantageous position with respect to security of supply, this idea runs entirely counter to the whole concept of the regional center where international cooperation and control, essential for the safeguards aspect of the operation, must of necessity lead to a limited abrogation of purely national interests and national rights. Under what circumstances would the host country accept this abrogation?

Chapter 9

Storing Radioactive Wastes in Cooled Pebbles

W. Bennett Lewis

It is not very widely recognized that two quite distinct methods have been proposed for the safe management of radioactive wastes from nuclear fission. One seeks to avoid water, the other employs water for cooling to immobilize the radioactive atoms in solids. Both methods have been described in the technical literature and to some extent demonstrated.

The well-known method often associated with the Oak Ridge National Laboratory is to seek a dry repository in an ancient salt bed or other stable geological formation to which the wastes may be consigned in perpetuity (Parker et al., 1958; Struxness et al., 1959-63). The repository would be protected from the biosphere and in particular from access by anyone or any animal not involved in the disposal. In Alvin Weinberg's phrase this protection requires "eternal vigilance."

The second method is that practiced normally in cleaning up any radioactive spill. The process is a double one, first to collect, concentrate, confine, and shield as much of the radioactivity as practicable in some store, the boundary of which is monitored until radioactive decay has made that no longer necessary. (For fission wastes stored in cooled glass pebbles, that would rarely exceed 1000 years, even when the wastes contain normal residues of plutonium and other actinides.) Any significant radioactivity escaping through the boundary (for example in this case the pebble surface) would, by the second part of the process, be again collected, concentrated, and confined within another boundary, or in the original where appropriate. Radioactivity escaping in water or air at concentrations comparable with the normal natural background is not considered significant and may be allowed to disperse.

This paper examines and expounds a practical form of the second method, which is suggested as most appropriate for large-scale nuclear fuel reprocessing plants including multinational facilities.

OUTLINE OF THE SECOND METHOD INVOLVING GROUND WATER

The particular practical form of the second method is to make use of ground water kept at a standard acceptable for public drinking water. The ground water should preferably flow at a higher elevation than the mean water level in surrounding large expanses, whether rivers, lakes, or seas. The ground water flowing out would then be subject to dilution by a large factor by mixing with these surface waters.

High-level radioactive wastes are fixed in relatively insoluble and highly durable glass pebbles directly cooled by the ground water. A satisfactory small-scale experimental demonstration of the method has been in operation for more than fifteen years. For the last five years the levels of radioactivity in the ground water have been below the maximum permissible concentrations recommended by the International Commission on Radiological Protection (ICRP) for public drinking water (ICRP, 1964).

These levels of radioactivity have been measured from time to time. One method of representing the results is to suppose that the activity is being leached out by slow dissolution of the glass. On this basis it is taking about four years to remove a monatomic layer of the surface, and it would take over 1.6 million years to remove a 100-micron thickness, during which time the activity in the water would have become so low that it would not be measurable. Incidentally, plutonium would have decayed to uranium as observed in the Oklo phenomenon discussed later in this paper.

Supposing, however, in another case, that the glass were in some way unstable in the early years so that the radioactivity in the ground water rises above a level acceptable for drinking water, then the second stage of the double process could be set in action to collect the escaping activity in a suitable filter or ion-exchange bed, and then again concentrate and fix it in solid glass. (The normal, very low rate of movement of atoms, especially of heavy atoms, within stable solids at ambient surface temperatures is reviewed under the heading "Chart of Atomic Mobility in Solids," below).

At any time that the ground water is below the level of activity for drinking water it may be allowed to flow out from the site to mix with other surface drainage. A state, however, to be avoided is that the water evaporates outside the waste-management site leaving a concentrated radioactive deposit.

It will be noted that the method possesses flexibility for meeting and dealing with emergencies, and such circumstances are similar to those that are liable to occur with the natural radioactivities in the uranium and thorium radioactive series, where much larger total amounts may, however, be involved.

The demonstration referred to is continuing at the Chalk River Waste Management area in Ontario, and in addition to the specific reports listed it may be helpful to refer to a general text, "Management of Radioactive Wastes," by C.A. Mawson (1965).

Those familiar with the great power of the methods of study using radioactive tracers will appreciate the many orders of magnitude between the high-level radioactivity stored in the pebbles, the very small fraction cumulatively escaping to the ground water, and the still smaller fraction at any time in the local ground water, usually needing concentration by a large factor to allow the radioactivity to be counted.

A specific calculation has been made and published as "Radioactive Waste Management in the Long Term" (Lewis, 1972), which concerns the management of all the radioactivity that would arise in a period of two thousand years if fission of uranium-233 supplied all the energy used by a world of 15 billion population (about four times the present population), supplied at a rate of 50 kilowatts (thermal) per capita; that is, about five times the current North American energy supply per capita.

All the high-level radioactive wastes would be stored in the glass pebbles. These wastes could also include the small fraction (up to 1 percent) of the uranium-233 and higher isotopes that remain with the waste from the reprocessing and recycling operations made after obtaining from the thorium-uranium fuel between 2 and 2.5 fissions per initial fissile atom in the fuel (that is, 2 to 2.5 megawatt days [thermal] per gram of the initial fissile material).

Even if spent fuel is processed to recover and recycle the fissile content within a year or two of removing the fuel from the reactor, it seems both practical and wise to delay the fixation of the wastes in a solid glass or ceramic form for about ten years. The advantage is that less cooling of the solid is needed because the decay heat will have diminished and also the chemical changes accompanying radioactive decay of the fission products will mostly have been completed.

The problem introduced by the presence of tritium is considered. It would become necessary to recycle the process water. Nevertheless some tritiated water may escape to contaminate the local ground water and require that also to be recycled. To maintain the essential cooling it is acceptable to operate the glass pebble storage at a slightly higher temperature and arrange for cooling by secondary pumped water taking heat from the primary water and then losing it by evaporation and dispersal. Although recycle of ground water would remove some of the simplicity from the system, it would serve also to diminish the total radioactivity eventually dispersed.

THE SMALL-SCALE DEMONSTRATION

The following information is derived from two recent papers by W.F. Merritt (1976; 1976a), "The Leaching of Radioactivity from Highly Radioactive Glass Blocks Buried Below the Water Table: Fifteen Years of Results" and "High-Level Waste Glass: Field Leach Test", as well as the references given to earlier publications including Watson et al., 1958 & 1960 and ICRP Publication No. 6 (1964).

The material chosen for the glass is 85 percent nepheline syenite, with a composition approximately $(Na_{0.75}K_{0.25})\ AlSi_2O_6$ with 15 percent lime (CaO). The mixture was combined with a nitric acid solution of fission products. The resulting gel was dried in crucibles, denitrated at 900° C, and then fused at 1350° C.

Twenty-five glass blocks, roughly hemispheres of 14-cm diameter initially containing 1100 Ci of six-year-old fission products, with a total surface area of $1.1 \times 10^4\ cm^2$ (equivalent to 25 spheres of 11.8 cm diameter) were buried in May 1960 below the water table in the Chalk River Waste Management Area discussed by Mawson (1965). The glass contains strontium-90 at 0.01 Ci/g and the ground water flows past at 39 l/day, and in 1974 carried away 1.2×10^{-4} μCi/l.

The ICRP (1964) discussed the $(MPC)_W$ for strontium-90 and set this for a full 168-hour week at 4×10^{-3} μCi/l corresponding to a body burden of 2 μCi in bone as the critical organ. Applying a factor of ten reduction for exposure of the general public, the maximum permissible concentration is 4×10^{-4} μCi/l, which is several times higher than the ground water. For cautions concerning the application of $(MPC)_w$, see Mawson (1965).

The ground water collects in a small lake where it is further diluted and from which a creek flows into the Ottawa River. The principle of management is to ensure that the radioactivity in the water in the creek is below the maximum allowable in public drinking water, and this condition has always been met.

For any particular radioactive waste fixed in glass pebbles it would be necessary to ensure over an initial period of perhaps ten years that the leaching rate is satisfactory for the overall water system, so that any water dispersed to the environment meets the drinking water standard and does not evaporate before sufficient further dilution and dispersal.

PLUTONIUM-239 AND STRONTIUM-90 AT LOW LEVELS IN THE ENVIRONMENT

It has been estimated that some 6 tonnes (0.3 MCi) of plutonium-239 and -240 has been dispersed via the atmosphere from the nuclear weapons tests that have already taken place (Hardy, Krey, & Volchok, 1973; Wachholz, 1974).

It has proved convenient to study together low levels of plutonium and strontium-90 resulting from the fallout from the weapons tests. An international symposium in November 1975 brought up to date the results from such studies, which were also extended to include cesium-137 and Fe-55. The proceedings of the symposium, which include significant discussion as well as the papers presented, have been published by the International Atomic Energy Agency, under the title "Transuranium Nuclides in the Environment" (IAEA, 1976).

When measuring plutonium in the environment it is not easy to analyze separately for plutonium-239 and plutonium-240. The alpha-rays they emit have much the same energy, so it is usual to record the combined activity in a given mass or volume. Geochemical and biological processes may cause some redistribution, especially in surface deposits and at other interfaces, but the trend of the bulk is to deposit on sediments that settle under gravity. In lakes, seas, and oceans it is observed that plutonium sinks more rapidly than strontium-90, and that is advantageous because when both fall out together from the atmosphere a general picture is most readily obtained by studying strontium-90, knowing that plutonium will be removed more rapidly from the biosphere.

In the fallout from weapons tests the activity ratio of plutonium-239 and -240 to strontium-90 is about 0.016, but in surface waters it is much lower, typically 0.003 (IAEA, 1976: 99). Moreover the $(MPC)_w$ for plutonium in drinking water is about 5×10^{-3} μCi/l, which is an order of magnitude higher than that for strontium-90. The biological actions of these substances are quite different, but it remains true that for the first few hundred years the toxicity of strontium-90 in fallout is greater.

The current level of concentration of plutonium in lake and river water is far lower than that considered the maximum permissible for drinking water. Moreover, if no further nuclear explosions in the atmosphere take place, the plutonium concentration in water will decay much more rapidly than the radioactive decay rate because of the propensity of plutonium to be absorbed by sediments and thus deposit on river, lake, or sea bottoms.

RADIUM AND ITS DAUGHTERS

Because of the different radiobiological action of nuclides emitting alpha-rays from that of beta- and gamma-emitters, it seems prudent to consider alpha-emitters separately in the environment.

The alpha activity of background plutonium discussed above is generally much below that of radium and its daughters, including polonium naturally occurring in rocks, sands, and soils. Moreover, mainly through the escape of radon, a heavy noble gas, the atmosphere, water, and sediments have a measurable alpha activity.

In many modern buildings with bare concrete walls the background radiation level increases to several times its normal, very low level when the ventilation is switched off. This happens because from many gravels and crushed-rock materials there is a natural emanation of radon, which has a radioactive half-life of 3.8 days. It leaves an atom of radium-A, now identified as polonium-218, that appears usually as a positive ion that attaches itself to a dust particle, but its half-life is only 3 minutes and so the chain goes on through 27-minute lead-214 (RaB); 20-minute bismuth-214 (RaC); to 22-year lead-210 (RaD). Lead-210 is an isotope we should take note of because of its 22-year half-life; and attention that should be paid its daughters, 5-day bismuth-210 (RaE) and 138-

day polonium-210 (RaF) (which is very apt to become airborne by aggregate recoil before it settles down as the stable lead-206). From a practical point of view we wish to keep both the short-lived active deposit radium-A + B + C and the wandering polonium-210 out of the air in low-background physics laboratories and in mines where higher levels up to those of significance to long-term health can arise. Good ventilation is the means available and used to keep the levels low enough.

When we wish to obtain a sufficient concentration of radon to study, we do not use gravels or crushed stone as a working source because so little of the radon escapes or emanates; instead we extract the parent radium into solution where, by radiolysis, hydrogen and oxygen are released and carry off with them the radon as it is produced.

THORIANITE

The low rate of escape of radon from gravels and crushed rock may be appreciated from observations of other gases and in particular of helium. Retention of substances in minerals over tens and even hundreds of millions of years came particularly to the notice of the scientific world in the early years of this century with the discovery, analysis, and study of helium in thorianite (Dunstan, 1904, et al.).

Pebble-sized broken fragments of this mineral were shipped from its source in Ceylon to London where it was found to contain helium at a very high concentration, typically 10 cm^3/g, equivalent to that in the gas at 100 atmospheres pressure. The origin of this helium was certainly the radioactive decay of uranium, thorium, and their daughter products. The question was, Why had it not escaped over the tens or even hundreds of millions of years since it was formed? It was finally realized that the natural rate of escape at ambient temperatures was in fact just so slow.

Thorianite could be heated to 400°C, and still most helium remained contained. But on heating still higher, to 1000°C, it would explode. These phenomena are now incorporated into the body of scientific knowledge and well understood; that is, consistent with the general understanding of the migration of atoms in solids. When we wish to store plutonium or fission products for thousands or millions of years, this is the field of understanding we bring to bear. Besides properly embedding the material in a solid, it is necessary to keep the solid cool.

CHART OF ATOMIC MOBILITY IN SOLIDS

The long retention of helium in thorianite can be related to the immobility of plutonium in cooled solids by a general chart of atomic mobility that summarizes the results of typical measurements. From this chart it may be inferred that a plutonium atom fixed in glass would not, even at 100°C, make—on the average—even a single jump to a neighboring position in the atomic lattice (≈(3Å, or 3×10^{-10}m, or 0.3nm) in a million years if it lived that long.

Since the chart (see Figure 9-1) is very general, it appears at first sight rather complex. The scale at the extreme right gives the lattice point dwell time. The figure of a million years (10^6 yr) just used comes near the bottom of the chart, which has been arbitrarily cut off at 30 million years.

Each of the radiating straight lines corresponds to a specific activation energy for atoms to jump from one lattice point to a neighbour. The lines are marked with these energies in electron volts (eV). For plutonium in glass the activation energy would be at least 2 eV.

The chart may be used as an approximate ready reckoner. This type of plot was originated a century ago by Arrhenius. His idea was that points derived from measured values of diffusion coefficients would fall on straight lines, the slope of which would be a measure of a characteristic activation

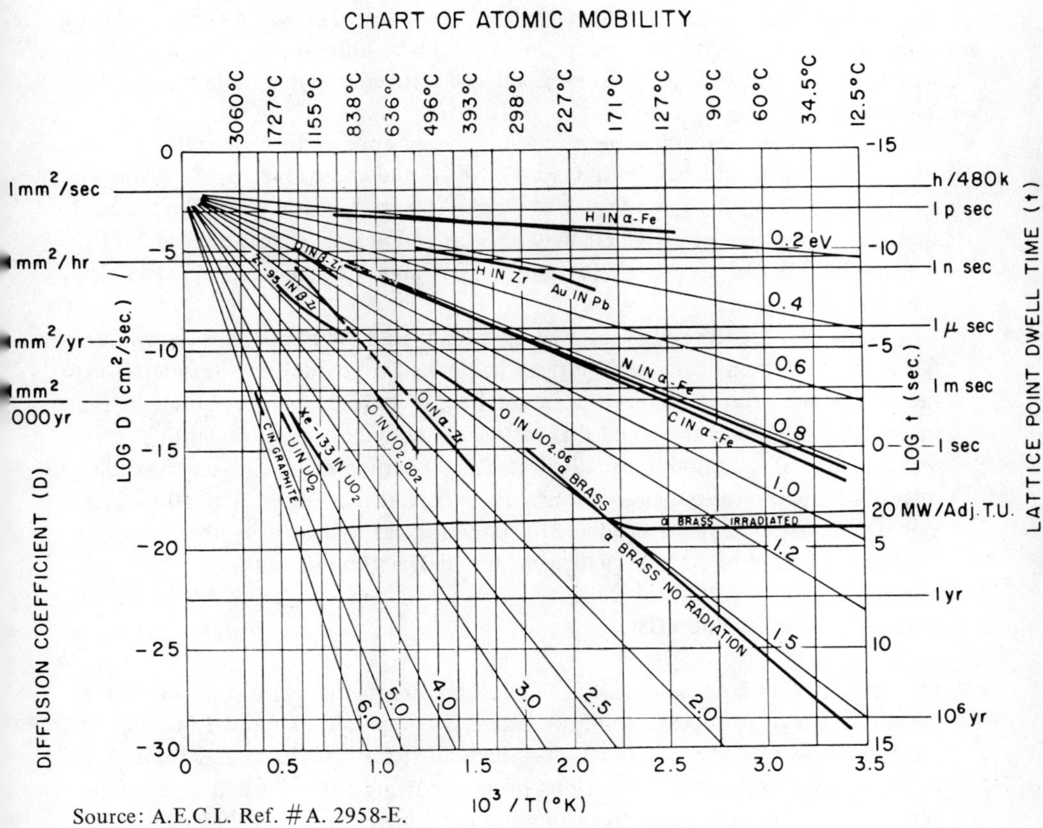

Source: A.E.C.L. Ref. #A. 2958-E.

Figure 9-1. Chart of Atomic Mobility.

energy. The heavy lines are measured values for particular atomic species in particular solids. Although the results trend towards the radiating lines, there are significant departures caused by phase changes and geometrical changes as well as activation by intense radiation.

The chart is in essence a plot of diffusion coefficients against the reciprocal of the absolute temperatures 1/T (Kelvin). The scales are derived by first noting that 10^{-15} cm^2, that is, $(3Å)^2$ is approximately the area of a single cell of a typical atomic lattice in solid materials. Dividing D (the diffusion coefficient) by 10^{-15} cm^2 then gives the frequency with which an atom makes a jump from one lattice cell to the next, or the number of jumps per second. Stated in terms of atomic constants, 10^{-13} sec = $h/480k$ where h and k are Planck's and Boltzmann's constants and 480 Kelvin is the Debye temperature for iron representing atoms of medium mass. Further explanation of the chart and more references are given in the original publication (Lewis, 1962).

It may be of interest to note that a heavy atom, such as plutonium, at temperatures below, say, 300°C in a silicate glass, would have an activation energy greater than 2 eV (greater than O in UO_2), but less than 4.3 eV (U in UO_2). Even at 2 eV and 100°C the jump time would be 1 million years. So plutonium would not on the average make even a single lattice jump in a million years if it lived that long.

For helium in thorianite the activation energy appears to be about 1 eV, so the jump time would be about 1 sec at 100°C and the diffusion rate given by log D (cm^2/sec) = –15, or D = 10^{-15} cm^2/sec $\approx$ 1 mm^2/300,000 yr, and a thousand times less at 30°C. (It may also be of interest to note that at 300°C the dwell time for hydrogen in zirconium is about 1 nanosec, which is also that for gold in lead.)

Despite the high mobility of hydrogen in metals, it is possible to engineer for long-life systems because hydrogen is blocked by oxide films on metal surfaces. The temperature limit is then set by the stability of these films or the mobility of oxygen in the metal that may allow temperatures up to 600°C.

An effect of radiation is also illustrated by experimental results for the order-disorder transition in α-brass. It might be expected that higher activation energies would be similarly overridden by radiation, as indicated by the line for a reactor irradiation of 20 MW(t)/adjacent tonne of natural uranium.

THE OKLO PHENOMENON

Elaborate studies are being made at laboratories across the world connected with the migration of isotopes in solid-ore bodies over a period of about 1740 million years at Oklo in Gabon. The isotopes include fission products and those produced by neutron capture, as well as those occurring naturally and from radioactive decay. Interest arose from the discovery in recent years that there were marked variations in the U-235/U-238 isotope ratio in uranium from these ores. It has been deduced that about 1740 million years ago, over a period possibly as long as 600,000 years, fission chain-reactions occurred locally in those ores, where the concentration of uranium and water set up the necessary conditions. At that time the U-235/U-238 isotope ratio in natural uranium would have been several percent, and small volumes of less than a cubic meter appear to have sustained a chain reaction until the buildup of fission products and the expulsion of water brought the reaction to a halt, leaving its imprint preserved in fascinating detail.

What is known about the slow migration of atoms in solids is used in the analysis, making the record intelligible and at the same time extending the knowledge. At the time of the reaction the flow of water and the temperature are not well established, but it is clear that the more volatile and soluble fission products escaped to some extent from the reaction zones. On the other hand, the plutonium, rare earths, uranium thorium, and bismuth stayed together within a few microns. Some movement of lead is suspected but the interpretation is complex. The information has been reviewed and to some extent developed by those interested in the prospect of permanent disposal of fission products and actinides in geological formations. The very limited migration may also prove significant in establishing confidence in the prospective storage in solid pebbles.

Quoting in translation from an original review in French by C. Frejacques et al. (1975): Several tonnes of plutonium-239 have been formed in the reactor(s); one measures in effect a rate of conversion to plutonium of about 0.45 to 0.50 (per fission). The greater part of the plutonium returned (by radioactive decay) to uranium-235 during the reaction which lasted about 500,000 years. For a neutron fluence of 10^{21} n/cm^2 it is estimated that for 100 atoms of plutonium-239 formed, 93 have decayed to uranium-235, 5 were fissioned and 2 went to plutonium-240 by neutron capture. The hypothesis that the plutonium was not mobile is well confirmed by detailed analysis of the correlation between fluences and depletions measured on traverses through the reaction zone; no discontinuities or irregularities are observed that would be attributed to a migration of the plutonium. The stability of the plutonium is further confirmed by the fact that as yet no sample enriched in uranium-235 has been found at the edge or outside the reaction zones. This holds down to the micron scale; an examination by an ion probe of the location of uranium-235 and -238 shows in effect perfect concordance of the two images that would not be expected if plutonium-239 migrated before its decay to uranium-235. Moreover this is strongly confirmed by the location of the fission products, but this deduction assumes the uranium has not aggregated and equilibrated after the reaction.

Thorium-232 results from the decay of uranium-236 ($t_{1/2} = 23.9 \times 10^6$y) which forms during the reaction by neutron capture in uranium-235 and effectively at the same time by decay of plutonium-240 ($t_{1/2}$ = 6600 yr), which is also a neutron capture product. Good agreement was found between the observed and expected concentrations of thorium in the mineral.

Somewhat similarly, bismuth-209 results from the decay of uranium-237 formed both by neutron capture in uranium-236 and from an (n,2n) reaction in uranium-238. It is also formed by decay of plutonium-241, but the amount by this route would be too small to be distinguished. Good agreement was found between the observed and expected concentrations of bismuth in the mineral. The precision of these latter agreements is, however, insufficient to have detected any but a large migration of plutonium.

As is to be expected in unravelling so complex a phenomenon, the course of events is not uniquely resolved. It is believed that the uranium ore was forming by deposition from water flowing through. The extent to which fresh uranium was being added during the reaction and what products were being carried away at different times remains uncertain. R. Naudet, who has taken the responsibility of coordinating the distribution of samples for analysis and has taken an interest in all aspects of the exploration and interpretation, concludes that the expulsion of water from the reaction zones provided the main control on the chain reaction. These uncertainties, however, detract very little from the detailed record so well preserved over the subsequent hundreds of millions of years.

WASTE MANAGEMENT AND RECYCLING

The idea of storing high-level radioactive wastes in glass blocks or pebbles has now gained fairly general acceptance, but there are different ideas about the disposal of the blocks. A simple method of storage in ground water below the water table has been studied, in which the ground water serves to keep the pebbles cool so that any escape of radioactive materials is slow. A small-scale experimental demonstration is proving satisfactory over fifteen years. The basic method is flexible, allowing recycle of the ground water if desired. This is not impractical, because the essential cooling can be achieved by evaporation. The direct cooling water may be evaporated when the level of tritium in the water is low. But secondary water following heat exchange with the primary water may be evaporated instead if the primary water contains too much tritium.

It is natural to entertain ideas of retaining all components of fuel assemblies at the processing plant for recycle when that becomes economic. Quite long storage times, such as twenty to fifty years, may be optimum before recycle. The nonfissile heavy elements–uranium-238 and thorium-232–that form the greatest mass in the residues are likely to be contaminated with their more highly radioactive isotopes and decay products. Dealing with thorium-232 that is liable to be contaminated by thorium-228 and its highly energetic decay products is discussed by Lewis (1972), where it appears that storage for fifteen years or more before recycle would be desired after freeing it from uranium-232, the parent of thorium-228. Dealing with uranium-238 may be even more complex because there exists an attractive alternative source, namely the tailings from the uranium-235 enrichment operations, at a low market price. Moreover, if the irradiated material is contaminated with uranium-232, with its 74-year half-life, it might be desirable to keep it segregated for thousands of years or purge it from thorium-228 immediately before re-use. There are also other possible complications, such as the presence of excess ionium (thorium-230) which has an 80,000-year half-life and is the parent of radium-226 with its 3.8-day radon emanation.

Of the other components of fuel assemblies, it may be practical to re-use zirconium to some extent after a few years, although the irradiated material has unfortunately a somewhat higher neutron-capture cross-section. On the other hand, nickel, which is still considered a possible component of stainless steel in fast-reactor fuel, is an element that can probably never be recycled because of the ingrowth of nickel-59 under irradiation. That nuclide not only has an 80,000-year half-life, decaying to cobalt-59, but also has a large cross-section for the $^{59}Ni(n,\alpha)^{56}Fe$ reaction. The helium produced in this reaction is structurally deleterious in the temperature range at which stainless steel is likely to be used in fast-neutron reactors.

It is clear that although the major tasks of radioactive waste management are to take care of the heat-producing fission products and the fissile nuclides, the chemical changes along the radioactive series complicate some recycling operations and impose timing considerations.

For the long term it seems necessary to plan for retrieval and recycling of major fuel components. Large-scale reprocessing plants with co-located waste-management facilities offer some conveniences and cost advantages.

REFERENCES

Dunstan, Wyndham. *Nature* 69 (1904): 510. Professor Dunstan discovered its composition and high helium content, detailed by W.R. Dunstan and G.S. Blake, *Proc. Roy. Soc.* A76 (1905): 253. M.W. Travers, pioneer with Sir Wm. Ramsay on noble gases and their analysis, arrived at the accepted explanation of helium release on heating, *Nature* 71 (1905): 248. E. Rutherford discusses, in "Some Cosmical Aspects of Radioactivity," *J. Roy. Astron. Soc. Canada,* May–June 1907(see collected papers, Vol. 1 [London: Allen & Unwin, 1962], p. 930). R.J. Strutt organized many studies of radioactive minerals and origin of helium, especially. *Proc. Roy. Soc.* A80 (1907/8): 572; and ibid., A82 (1909): 166; ibid., A84 (1910): 379; also D. Orson Wood, *Proc. Roy. Soc.* A84 (1910): 70, who studied helium release by action of heat and noted only small release at 400°C but explosion at bright red heat. More recently R.S. Barnes and D.J. Mazey, in *J. Nucl. Energy* 5 (1957): 1–3, studied its spontaneous fragmentation at 950°C on slow heating.

Hardy, E.P., Krey, P.W., and Volchok, H.L. "Global Inventory and Distribution of Fall-Out Plutonium," *Nature* 421 (1973):444.

IAEA. *Proceedings* of a Symposium on "Transuranium Nuclides in the Environment," 17–21 November 1975 at San Francisco, jointly organized by USERDA and IAEA. SM-199.

ICRP, 1964. International Commission on Radiological Protection, Publication No. 6, 1964.

ICRP, 1972. "The Metabolism of Compounds of Plutonium and other Actinides," International Commission on Radiological Protection, Publication No. 19, 1972.

Le Phenomene d'Oklo. Proceedings of Symposium of Libreville, SM/204, Gabon, 23–27 June 1975, IAEA, Vienna, 1976, STI/PUB/405. C. Frejacques et al., "Le plutonium-239," in ibid., pp. 509–24.

Lewis, W. Bennett, 1962. "Criteria for the Selection of Materials for Water-Cooled Reactors, with Comments on D_2O Reactors," *AECL-1567 and Nuclear Metallurgy VIII*. New York: AIME 1962. Based on reviews up to 1960 including: C.E. Birchenall, "The Mechanism of Diffusion in the Solid State," *Metallurgical Reviews* 3 (1958): 235; D. Lazarus, "Diffusion in Metals," *Solid State Physics*, Vol. 10 New York: Academic Press, 1960; W. Seith, "Diffusion in Metallen," Springer-Verlag, Berlin, 1955; A.B. Auskern and J. Belle, "Uranium Dioxide: Properties and Nuclear Applications," 1961, ed. J. Belle, USAEC; G. Dienes and A.C. Damask, "Radiation Enhanced Diffusion in Solids," *J. Appl. Phys.* 29 (1958): 1713.

Lewis, W. Bennett, 1972. "Radioactive Waste Management in the Long Term", DM-123, AECL-4268. CRNL, October 1972.

Mawson, C.A. *Management of Radioactive Wastes*. New York: Van Nostrand, 1965.

Merritt, W.F., 1976. "High-Level Waste Glass: Field Leach Test," 1976 Annual Meeting of Amer. Nucl. Soc., Toronto, 13–18 June 1976 (to be published). Abstract in *Trans. Amer. Nucl. Soc.* 23 (June 1976): 167–168.

Merritt, W.F., 1976a. "The Leaching of Radioactivity from Highly Radioactive Glass Blocks Buried Below the Water Table: Fifteen Years of Results," IAEA/NEA Symp. on the Management of Radioactive Wastes from the Nuclear Fuel Cycle, 22–26 March 1976, Vienna. Also as AECL-5317.

Merritt W.F., and Parsons, P.J. "The Safe Burial of High-Level Fission Product Solutions Incorporated into Glass," *Health Physics* 10 (1964): 655–664; W.F. Merritt, "Permanent Disposal by Burial of Highly Radioactive Wastes Incorporated into Glass," *Proc. Symp. Disposal of Radioactive Wastes into the Ground* (Vienna: IAEA,1967), pp. 403–408.

Parfenov, Yu. D. "Polonium-210 in the Environment and in the Human Organism," *Atomic Energy Review* 12, No. 1 (1974): 75–143.

Parker, F.L., et al. Report ORNL-2560, pp. 11–21, 1958; also Struxness, E.G., et al. Health Physics Division Annual Reports, ORNL-2806 (1959); -2994 (1960); -3189 (1961); -3347 (1962); -3492 (1963).

Wachholz, B.W., comp. USAEC Conf., "Plutonium and Other Trans-uranic Elements: Sources, Environmental Distribution and Biomedical Effects," Dec. 1974 (WASH-1359).

Watson, L.C.; Durham, R.W.; Erlebach W.E.; and Rae, H.K. "The Disposal of Fission Products in Glass," P/195, Proceedings 2nd U.N. International Conference on the Peaceful Uses of Atomic Energy, Vol. 18, 1958. Also, L.C. Watson, A.M. Aikin, and A.R. Bancroft, "The permanent disposal of highly radioactive wastes by incorporation into glass," *Proc. Conf. Radioactive Wastes,* Monaco, 1959, IAEA Vienna, 1960.

Chapter 10

Physical Security in Multinational Nuclear-Fuel-Cycle Operations

Mason Willrich

Physical security is an important consideration in most operations in the civilian nuclear fuel cycle downstream from uranium mining and milling. This is the case whether the activity involved is conducted by private or public enterprise, or under national or multinational governmental control. By "physical security" (the U.S. terminology), or "physical protection" (the IAEA terminology), we mean measures intended to deal with the possibility of theft or sabotage of nuclear materials or facilities.

This paper explores the question of physical security in civilian nuclear-fuel-cycle operations under various forms of multinational control. First, the security threats are outlined and the basic features of a physical security system are discussed. Next, physical security principles and requirements are assessed in the context of various forms of multinational control.

The paper is largely conceptual. At this writing physical security systems applicable to the civilian nuclear fuel cycle are undergoing intensive reevaluation in the United States, the focal point being the pending Nuclear Regulatory Commission (NRC) decision on widescale recycling of plutonium-bearing fuel in light-water reactors (LWR's) (US NRC, 1975). The NRC will reach a decision in mid-1977 at the earliest, and major changes in physical security requirements are probable in connection with an approval of plutonium recycling. The outcome of the U.S. decision-making process is likely to influence, though not determine, the future development of physical security systems in other countries.

Another reason for the largely conceptual nature of the discussion here is that, to the extent there has been relevant experience, it is not a matter of public record. Evidence of attempted or threatened acts of nuclear theft, sabotage, or terrorism and operational experience with physical security measures designed to cope with such acts have not been, and are not likely to be, fully disclosed to the public. Moreover, the detailed characteristics of physical security systems in operation in various specific plants are withheld from the public record so that potential troublemakers will not have a blueprint of the system they intend to overcome. Emphasis on either fragmentary news stories, which expose deficiencies, or government and industry claims of efficacy, may muddle clear thinking about the basic issues involved in physical security systems as they relate to the future worldwide growth of the nuclear power industry.

This paper's primary focus is on the "back end" of the fuel cycle–reprocessing, plutonium-bearing fuel fabrication, and associated irradiated fuel and plutonium storage and transportation. The main risks of nuclear theft exist in this part of the fuel cycle. In a nuclear power industry that includes plutonium recycling, this is where large quantities of plutonium will be present in a form in which it might be stolen and thereafter fabricated directly into crude nuclear explosives.

Presently and for the near future at least, the complexity and cost of uranium enrichment poses formidable technical obstacles to the possibility that criminal groups might engage in clandestine uranium enrichment. Stringent physical security measures must be taken, however, to prevent theft of highly enriched uranium prior to use as fuel in research reactors, high-temperature gas-cooled reactors (HTGR's), and, possibly, as start-up fuel in liquid-metal fast-breeder reactors (LMFBR's).*

RISKS OF THEFT AND SABOTAGE

The sources of risk of nuclear theft or sabotage are human beings (Willrich & Taylor, 1974:107-20). Any such action would be a criminal act. Therefore, in considering physical security measures it is useful to divide the sources of risk into two main categories: (1) employees and other persons who are authorized to have access to nuclear materials or facilities; and (2) persons not so authorized. A physical security system must be able to cope with both internal and external threats. Such a system must also cope with the possibility of a criminal group or conspiracy that includes both authorized and unauthorized persons.

One individual acting alone may attempt nuclear theft or sabotage, or a plot may be attempted by groups of various sizes and capabilities. The motive

*From an economic viewpoint, it would be highly desirable if a simple, cheap uranium enrichment method could be discovered. From a security viewpoint, however, such a technological breakthrough would appear to be destabilizing. Among other things, it would mean that criminal groups might obtain high-enriched uranium for use in nuclear explosives by conducting their own enrichment operations on a small scale. The discovery of a simple uranium enrichment method–one that is much simpler than any known techniques using diffusion, centrifugation, nozzle or laser processes–would thus give rise to serious physical security problems in uranium mining and milling operations.

may be monetary, political, irrational, or, perhaps more likely, a complex blend of various motivational elements.*

Without taking physical security into account, the probabilities of nuclear theft or sabotage attempts would depend largely on the strength of motivations for such criminal behavior within society. The consequences of a successful nuclear theft would depend largely on the intentions and capabilities of the criminals involved.

The consequences might include huge financial losses for the target enterprise through extortion. A nuclear theft might also lead to construction of an explosive and threatened or actual destruction of a political nerve center or urban complex with a nuclear explosion. The consequences of nuclear sabotage might include the destruction of the facility concerned (resulting in property damage of several hundred million dollars and substantial disruption of a vital industry), the death or injury of workers at the facility, and radiological damage to persons and the environment in the surrounding area.

If a successful nuclear theft should occur, it is by no means certain what the consequences would be. They might range from rather limited effects at one end of the spectrum through a series of more severe impacts. In the worst case, it is possible that the results would be catastrophic, but the probability that consequences at this end of the range will actually occur is quite small. As in other aspects of the nuclear enterprise, balancing this very small probability against the very large potential consequences presents one of the most difficult problems of social judgment.

Aside from physical security measures, the dominant factors determining the levels of risk of theft or sabotage at any time will be extrinsic to—and largely beyond the control of—the nucleaar power industry and the government agencies with primary responsibility for regulation of that industry. Nuclear power reactors and related fuel-cycle facilities embody very large and long-term commitments of financial and technical resources in particular geographic locations. The risks of theft or sabotage at any particular location are uncertain and likely to be volatile over the projected thirty-year operational lifetime of any facility in a specific location. In such a variable risk environment, the essential role of physical security is to assure that the probabilities of a successful act of nuclear theft or sabotage and of catastrophic consequences resulting from such an act are sufficiently low as to be socially acceptable.

SECURITY AGAINST THEFT OR SABOTAGE

Bearing in mind the manifold sources of risk, the potential for catastrophe, and the volatile characteristics of any projection of probabilities of human

*It is also possible that one or more persons involved in a nuclear theft or sabotage attempt might be acting under the compulsion of threatened harm to a member of his or her family previously kidnapped by the perpetrators of the attempt.

behavior decades into the future, we turn to the question of how to protect society against nuclear theft or sabotage attempts (Willrich & Taylor, 1974: 121-62). In so doing, we briefly consider physical security system objectives, functions, and measures. The industrial structure of the back end of the fuel cycle will also have an important bearing on the design and effectiveness of a physical security system, and these structural issues are considered.

Objectives

From the viewpoint of society, the objective of physical security is to protect the public against undue risks of death, personal injury, or property damage resulting from theft of nuclear materials or sabotage of facilities used in nuclear industry. From the viewpoint of nuclear industry, the objective is to protect industrial operations from undue risks of loss in terms of money, governmental permission and public acceptance.

What is an "acceptable" level of risk or "adequate" protection is essentially a political question. The answer is likely to vary widely among different national societies and change within one society over time. At any specific time and place, the answer will depend largely on perceptions of the social importance of nuclear power and of the threat to society of criminal behavior, especially nuclear terrorism.

The International Atomic Energy Agency (IAEA) has developed a set of recommendations for the physical protection of nuclear material (IAEA, 1975). These recommendations posit that the objectives of a state's physical protection system should be "to establish conditions which will minimize the possibilities for unauthorized removal of nuclear material or for sabotage," and "to provide information and technical assistance in support of rapid and comprehensive measures by the State to locate and recover missing nuclear material" (IAEA, 1975:¶21[a], [b]). A draft statement of the objective of the U.S. physical security system is that the system should "deter, prevent, or respond to (1) the unauthorized possession or use of significant quantities of nuclear materials through theft or diversion; and (2) sabotage of nuclear facilities. . . ." so as to "insure against significant increase in the overall risk of death, injury or property damage to the public from other causes beyond the control of the individual" (USAEC, 1974:V-12). This statement will be reviewed and probably revised in the course of the NRC's plutonium recycle decision.

The results of these attempts to articulate physical security objectives are so vague as to be almost useless. It is noteworthy, however, that both statements more or less relate physical security objectives to the functions of the system—deterrence, prevention, and response in the U.S. national case.

Functions

The functions of a physical security system may be broadly or narrowly defined. Broadly defined, physical security functions may include deterrence, prevention, detection, response to theft or sabotage attempts, recovery of previously stolen nuclear material, and response to threats possibly involving such material.

The central function of physical security is to *prevent* theft of strategic materials or sabotage of sensitive facilities used in the nuclear power industry. In order to accomplish this, any attempted act of theft or sabotage must first be detected prior to its completion and then frustrated by a timely and effective response. A physical security system that is perceived by those who might contemplate theft or sabotage as being capable of preventing a wide range of credible threats is a powerful deterrent to such criminal behavior.

For several reasons, prevention and deterrence have paramount importance. The chances of recovery of stolen material following a theft are likely to be slight. Widespread knowledge of the fact of a successful theft of nuclear material may itself raise public anxiety, even without a specific nuclear threat based on the stolen material. One successful theft or sabotage may provide psychological incentives for further attempts. Perhaps even more important, potentially catastrophic consequences may result from a single successful nuclear theft or act of sabotage.

In addition to prevention and deterrence, a physical security system must also detect a theft of nuclear materials that has been completed despite preventive measures. In this respect the main problem is detection of thefts of small quantities by persons authorized to have access to nuclear materials. A theft-detection capability provides an assurance that preventive measures are functioning properly and also a trigger for recovery operations in case preventive measures fail. Moreover, the existence of a detection system serves as a deterrent to employee thefts.

Finally, a physical security system may include, or is at least closely related to, the function of responding to nuclear threats involving or believed to involve stolen materials. Nuclear threats may, of course, occur in places and at times which are remote from the scene of the original material theft.

Physical security considerations thus penetrate deeply into both complex industry operations and sensitive police activities.

Measures

In developing a physical security system, a variety of measures may be combined together. The system may be designed to cover primarily deterrence, prevention, detection, and response to theft or sabotage attempts, or the system can be conceived at the outset with the functions of recovery and response to threats of nuclear violence also in view. The following discussion focuses on the main types of security measures applicable to the back end of the nuclear fuel cycle, including fuel reprocessing and fabrication plants and associated storage facilities and transport links.

Prevention of employee theft or sabotage. Measures to prevent employees of nuclear industry from stealing nuclear material or sabotaging facilities include security clearances, access limitations and inspections, and surveillance of work areas.

There are several reasons why security clearance of employees who are to have access to sensitive nuclear materials and equipment may be desirable. Prior clearance may reduce the risk of sabotage by disgruntled or unstable employees. Security clearances may also reduce the risk of employee theft of very small quantities of plutonium. Moreover, a link with the inside of a sensitive facility would be very valuable for any outside group planning theft or sabotage, and prior security clearance of employees in the most sensitive jobs may help to ensure that such links have not been forged.

However, requiring special clearance of employees may raise a number of labor-relations problems and may be a potential source of conflict between management and unions. A security clearance requirement may reduce the flexibility of management in making work assignments, and also may restrict job mobility of workers. Jobs may be denied on grounds that have nothing to do with merit or technical qualifications. Moreover, the denial of a clearance to an employee may prejudice his or her subsequent career anywhere in the nuclear power industry, even in less sensitive positions.

Whether or not special clearance is required as a precondition of employment, the fewer the persons who have access to sensitive nuclear material and equipment the better from a security viewpoint. Access to areas where such material or equipment is located may be limited to those whose jobs require it. Limitations may take the form of some combination of physical barriers and administrative controls.

The design of jobs in nuclear industry is thus important. In addition to establishing special controlled-access areas, it may be desirable from a security standpoint to separate jobs that require access from other jobs in the industry. But while facilitating control, such compartmentalization may adversely affect employee morale and productivity.

Security inspections at entrances and exits of material- or equipment-access areas may be among the most cost-effective employee surveillance measures. Of course, employees themselves and any parcels they are carrying may be searched with varying degrees of thoroughness. Special work clothes may also be required. Changing clothes may be required and also monitored. Special work clothes are often necessary for health purposes and, if so, the only additional security burden would be in carefully constructing the clothes-changing procedures. Exit doorway monitoring instruments may also be used.

There are also many possibilities for direct monitoring of the activities of employees who are authorized to have access to sensitive nuclear materials or equipment. At least two persons may be required to be present in sensitive work areas. Instead of (or in addition to) a two-man rule, employees engaged in sensitive work may be kept under continuous remote surveillance by the security staff at the facility, perhaps by closed-circuit television monitoring.

Prevention of theft or sabotage by outsiders. Various containment measures may provide security against theft or sabotage by individuals or groups outside the nuclear industry. Such measures include heavy containers, massive vaults, barriers such as thick walls and high fences, locks, alarms, remote surveillance, automatically activated weapon systems, unarmed watchmen, armed guards, law enforcement authorities, and military forces.

Three factors are important to bear in mind in considering these various components of a physical security system:

1. The overall effectiveness of measures to prevent theft or sabotage will be determined largely by the combined effectiveness of the physical containment structures installed and the human security forces deployed or available. A strong physical barrier may retard completion of a theft or sabotage and reduce the need to use force immediately. Furthermore, alarms, physical barriers, and on-site security forces may, together, ensure sufficient time for off-site forces to stop a theft or sabotage in progress.
2. Physical barriers alone cannot provide adequate protection. Given enough time, a group of men equipped with modern tools are likely to be able to penetrate any physical barriers that could be erected at a reasonable cost. So, the capability to use force under human direction and control must be an integral part of a physical security system.
3. The resistance of a physical barrier to a given amount or force is somewhat predictable. However, the extent of resistance one person will use to oppose force threatened or used by another may be quite uncertain.

The requirement for a capability to meet force with deadly force raises two primary issues: What postulated threats should the system be designed to defeat? And who should have responsibility for the use of force, if necessary to prevent theft or sabotage?

Credible external threats range from one individual acting alone to a well-armed group of substantial size, say six to ten persons. The difficulty of dealing with the upper range of threats is compounded by the fact that a group with the capability to use substantial force in the execution of a nuclear theft appears likely to attempt to steal a large quantity of plutonium. In some countries, nuclear theft or sabotage might be part of a larger conspiracy to overthrow the government in power.

In analyzing the capabilities of a security force to cope with such a range of threats, on-site and off-site forces may be considered as a whole. If they are, the means available to bring any off-site forces to bear become critical to the effective functioning of the system as a whole. Unarmed watchmen cannot be expected to deter a determined criminal group. However, the capabilities of on-site armed guards to do much more than unarmed watchmen should not be overestimated. A watchman does not become an effective armed guard by hanging a holster with a loaded pistol on his belt.

The capabilities of off-site law enforcement authorities to deal with nuclear theft or sabotage attempts should also be carefully examined. Communication between the on-site security force and off-site assistance may be continuous or sporadic. Nuclear reprocessing facilities are likely to be located in rural, sparsely populated areas. Law enforcement authorities in these areas, even if they are willing to cooperate, may lack the manpower, equipment, and training to deal with the maximum credible threat at the facility. This suggests that it may be desirable in some locations to have on-site the greater part of whatever security forces are deemed necessary.

Extensive training and preparation of the forces in all these categories is essential. Much of this is similar to what is needed for operational personnel. It should include instruction in the physical and other properties of radioactive materials, a thorough familiarity with the plant layout and operations, as well as frequent drills and simulation exercises. Unless this is carefully done in advance, response in an emergency is likely to be confused and ineffective.

The preceding analysis leads to the question of who should be responsible for developing and maintaining the required security force or forces. The enterprise itself is one alternative, and the government under whose jurisdiction the enterprise operates is another. Both advance coordination and on-the-spot cooperation are necessary if the responsibility for preventing nuclear theft or sabotage is divided among the nuclear enterprise and law enforcement authorities. In any event, when a threat occurs, orders will come from immediate superiors and critical actions will be taken on the initiative of supervisors at the scene of the crime.

Under criminal law, the use of force–including the taking of human life–is justifiable in self-defense and in defense of other persons. In certain limited circumstances, the use of deadly force may be justified in the defense of property. The use of deadly force to cope with a nuclear theft or sabotage attempt would seem to be justifiable in view of the potentially catastrophic consequences of a successful completion. However, the precise legal consequences of the more extreme forms of action to prevent nuclear theft or sabotage attempts are likely to vary considerably from one country to another.

In general, it would be destablizing to society as a whole if citizens viewed self-help as their *primary* means of protection against crime. Accordingly, it would seem appropriate for the security forces required for physical protection of nuclear material and facilities to be provided by the government rather than the enterprise concerned. This view is, however, counter to the current trend of thinking within the U.S. government.

Industry Structure

Physical security will be affected not only by the measures applied to sensitive nuclear material and equipment, but also by specific industrial development decisions and operating practices. But physical security is unlikely to be the dominant factor in such matters. Three areas of industry operation have particular significance.

Reprocessing and plutonium recycling. At present the nuclear power industry faces a wide range of choices regarding reprocessing and plutonium utilization:

–Large-scale reprocessing may be delayed until the commercial attractiveness of breeder reactors is established, and irradiated fuel elements may be stored in the interim. Reprocessing would be limited to amounts required for breeder-reactor development and demonstration programs, and such reprocessing would be confined to the few countries with substantial breeder programs.

If adopted, this option would severely limit the amount and location of plutonium available in a form amenable for use in nuclear explosives in the near future, thereby substantially reducing the physical security problem. It would also delay the need for back-end multinational nuclear fuel centers. Multinational irradiated fuel storage facilities would be possible, but the incentives for such facilities would appear to be low.

–Large-scale reprocessing may proceed in order to recover fuel materials for recycling in light-water reactors. This option involves several sub-options.

–Only uranium may be recovered, leaving plutonium and radioactive wastes unseparated. The risk of plutonium theft would be minimal, but for economic reasons it seems very unlikely that this option will be adopted.

–Uranium and plutonium may both be recovered. If this step is taken, several additional options regarding disposition of the plutonium emerge. The recovered plutonium may be: (1) converted from nitrate to oxide; (2) blended and diluted with uranium oxide; (3) spiked with various radioactive materials; and (4) either stored or fabricated into fuel and recycled in light-water reactors.

From a security viewpoint, once plutonium has been recovered, the quicker it is diluted and recycled, the better. While spiking may increase the hazards of plutonium to a thief, it would also increase the health and safety risks of workers. For this reason it does not seem to be worthwhile pursuing, although various types of blending or dilution are being seriously considered by industry as a useful measure to reduce security risks in plutonium management.

Plant size. The benefits of economies of scale may be realized by the construction of large reprocessing and plutonium-bearing, mixed oxide fuel fabrication plants. From an economic viewpoint, the optimum plant sizes appear to be about 1500 tonnes for a reprocessing plant, complemented by 300 tonnes for a mixed oxide fuel fabrication plant. Such large plants would each be capable of servicing about 50,000 megawatts of generating capacity. However, much smaller reprocessing and fuel-fabrication plants may be built, especially in the earlier stages of nuclear power development. Small fuel-cycle plants may be desired as a means to gain industrial experience.

It is important to emphasize that, if available economies of scale are fully realized in nuclear industry operation, the industry is much more likely to develop along lines that are compatible with efforts to prevent or retard the spread of nuclear weapons than if large numbers of small, pilot-scale plants are constructed.

Facility location. In addition to being conducted in plants of various sizes, each discrete operation in the back-end of the nuclear fuel cycle may be carried out in more or less widely dispersed locations. For example, when plutonium is initially recovered it is in a liquid nitrate solution. It may be shipped to a separate plant for conversion to oxide, shipped again for blending or pelletization, shipped again for fuel-element fabrication, and finally shipped to a reactor for irradiation. Between each point the plutonium may be stored in small or large amounts for short or long periods of time. Alternatively, it is feasible and economical to perform reprocessing, conversion, and fuel-fabrication operations in one location, keeping inventories of recovered plutonium at minimum working levels.

From both economic and security viewpoints, the fewer the locations of back-end fuel-cycle facilities, the better. Particularly from the viewpoint of physical security, the co-location of reprocessing and fuel-fabrication facilities would eliminate the transportation link in the nuclear fuel cycle that appears to be the most vulnerable to plutonium theft.

Concentration of nuclear-fuel-cycle operations in the fewest possible locations might also enable safer sites to be chosen. Finally, if the proliferation of nuclear weapons is to be avoided, multinational reprocessing plants make little sense if the recovered plutonium is thereafter sent to a variety of other locations for fuel fabrication.

PHYSICAL SECURITY IN MULTINATIONAL NUCLEAR-FUEL-CYCLE OPERATIONS

Having considered physical security against nuclear theft and sabotage from a generic viewpoint, we are now in a position to discuss the possible effects of multinationalization on the objectives, functions, and measures that may be incorporated into a physical security system. Before embarking on this discussion, however, it is important to recall the many meanings of "multinational."

The participants in a multinational nuclear enterprise may include several countries, or only two. The countries involved may be in the same or different geographic regions. The participants may include state-owned, private, or mixed enterprises. Possible forms of multinational participation include financial, managerial, and operational. Discrete operations in the back end of the nuclear fuel cycle which might be multinationalized include irradiated fuel storage, fuel reprocessing, plutonium-bearing fuel fabrication, and radioactive waste storage.

In each case, of course, physical security arrangements would be largely affected by the identity of the participants, the structure of participation provided, and the discrete fuel-cycle operations involved in a multinational nuclear enterprise. Space and time do not permit consideration of every variation here, so the discussion is limited to the major ramifications.

Allocation of Authority

How would multinationalization of a nuclear-fuel-cycle operation affect physical security in that operation? The threshold inquiry in addressing this question concerns allocation of authority. In general, the authority to assure physical security in a nuclear-fuel-cycle operation is vested initially and primarily in the government under whose jurisdiction the activity is conducted. In the case of an operation conducted by private enterprise, the government concerned may impose physical security requirements on the enterprise and monitor and enforce compliance. Alternatively, the government may retain direct responsibility for certain aspects of the security system; for example, a guard force. The national government may also share authority with political subdivisions.

The participants in a multinational nuclear enterprise would initially consent to, or implicitly acquiesce in, the physical security measures to be applied. Moreover, the host government, on whose territory the multinational facility would be constructed, would have a special interest in the matter.

There are four broad alternatives for allocating physical security authority over a multinational nuclear enterprise: (1) authority may be allocated exclusively to the host government; (2) authority may be allocated exclusively to the multinational enterprise; (3) authority may be shared in various ways between the host government and the enterprise; or (4) authority may be shared in various ways between the host government, the enterprise, and an international organization such as the IAEA.

Objectives.

The objective of physical security is to reduce the risks of theft or sabotage to an acceptable level. Multinationalization may have various effects on this objective. The participants generally, or the host government in particular, may perceive the risks of theft or sabotage in a multinational nuclear-fuel-cycle operation as lower, higher, or the same as in a national or private enterprise.

If, for example, a multinational enterprise includes plant employees from a variety of countries, some participants may perceive the risks of employee theft or sabotage as reduced, since the employees of diverse nationalities would have incentives to check on each other. The host government may agree with such a risk assessment, or it may conclude that the risk of employee theft is raised by the greater likelihood that a foreign employee would be a criminal than one of its own nationals.

On the other hand, some participants in a multinational nuclear enterprise may be concerned that the host government would itself concoct and implement a nuclear-theft scenario at the multinational facility on its territory as a means of obtaining nuclear material for its own weapons program. If such fears exist among the participants in a possible multinational nuclear enterprise, they can be expected to surface in the choice of location of the facilities involved. Of course, the facilities should be located in a country where the host government is deemed trustworthy by all the participants.

It seems likely that various participants will perceive somewhat differently the level of security risk in a multinational nuclear enterprise. Nevertheless, each participant must, implicitly if not explicitly, conclude that—in light of the total circumstances—the level of risk of theft and sabotage is acceptable to it before the enterprise can be launched with that participant as a member.

In addition to the question of the effect of multinationalization on risk perceptions, there is the question, What is an acceptable level of risk? Would a host government accept a higher level of security risk in a multinational enterprise than in a domestic nuclear enterprise? One host government may strongly desire the enterprise to be located on its territory, and be willing to incur some additional security risk to obtain it. A different host government may insist on security measures in a multinational operation that are at least as stringent as those applicable to national or private nuclear operations subject to its jurisdiction.

Would the other participants in a multinational enterprise be likely to insist on a lower level of risk and hence more stringent physical security arrangements than the host government desires? If the multinational enterprise is to be located in a country with a large existing nuclear industry, it may not be reasonable for the other participants to insist on a higher level of physical security in a multinational operation than in comparable domestic operations. However, the multinational enterprise may be unique in the host country. In such a case, other participants may have reason to insist on a lower level of security risk than the host government would otherwise apply.

Functions

As discussed previously, the possible range of functions of a physical security system for the nuclear power industry include deterrence, prevention, detection, and response to attempted theft or sabotage, as well as recovery of stolen materials and response to threats involving previously stolen material.

If physical security authority is allocated generally to the host government, multinationalization will raise no special problem about the scope of functions involved. However, if security authority is divided or shared in some way between the host government and the multinational enterprise or an international organization, then careful functional analysis is necessary.

Certain physical security functions are more likely to be allocated to a multinational enterprise than others. The enterprise itself would of course, be responsible for the design, construction, and normal operation of the fuel-cycle facilities involved in accordance with physical security regulations, whether those regulations are self-imposed by the enterprise management or required by the host government.

Would a host government agree to allocate authority to a multinational enterprise to respond actively to theft or sabotage attempts or recovery of stolen material? Substantial sharing of authority in these functional areas seems unlikely for a variety of reasons. An active response to a theft or sabotage attempt may involve the threat or use of deadly force. The attempt would be criminal under the laws of the host country, and host-country citizens might well be involved. Furthermore, the consequences of a failure to respond effectively to a theft or sabotage attempt are likely to be felt primarily in the host country. These factors may lead the host government to insist on exclusive control of any response to a nuclear theft or sabotage attempt aimed at a multinational nuclear enterprise located within its jurisdiction. Moreover, the other participants in such a multinational enterprise are likely to be reluctant to intrude deeply into matters of such fundamental concern to the host government itself.

What has been said above about responses to theft and sabotage attempts applies even more strongly to recovery of stolen material and responses to threats involving previously stolen material. These physical security functions involve not only police activities, but also intelligence operations. Outside governments may be quite willing to cooperate, but responsibility for the conduct of such operations is likely to remain exclusively that of the host government, regardless of the multinational character of the nuclear enterprise involved. Therefore, the functional areas within which a multinational enterprise might have some authority for physical security are likely to be restricted to those which can be implemented through measures applied at the site of the facility, with the emphasis being on implementation of relatively passive or routine measures.

A division of physical security authority between the host government and a multinational nuclear enterprise may raise problems at the interface. These problems are similar to those which exist in a country where nuclear industry is conducted by private enterprise. The fact that a nuclear enterprise is multinational may raise somewhat more difficult problems in relating the enterprise effectively to local police and to national intelligence authorities, however.

It seems unlikely that the participants in a multinational nuclear enterprise would allocate substantive authority for any physical security function to an international organization such as the IAEA. The IAEA role is likely to be confined to a forum for exchange of physical security information, and possibly the development of standards. Of course, some indirect IAEA monitoring of certain features of a state's physical security system would be possible in the course of the application of the agency's materials accountancy safeguards.

Security Measures

In considering the effect of multinationalization, as in the previous generic discussion of physical security measures, it is useful to discuss measures designed primarily to deal with internal threats separately from measures to cope with external threats.

Security clearances, access limitations, and surveillance of work areas. As discussed above, these are the main measures to deter and prevent theft or sabotage by employees or other persons authorized to have access to sensitive nuclear materials or equipment.

The nature of the security clearance problem in a multinational nuclear enterprise would depend largely on the nationality of the employees who are to have authorized access to sensitive materials and equipment. Such employees may be provided only by the host government, or they may be hired by the enterprise from various participating countries (or elsewhere) according to expertise or national quotas, or they may be hired by the enterprise, subject to approval of the host government.

If the key employees must be nationals of the host government, then presumably any security clearance will be that of the host government alone. This implies that the other participants trust the host government not to use the employees it provides to steal from or sabotage the multinational enterprise.

If the key employees are to be multinationally recruited, a number of questions will arise about any security clearance requirement. Each participating government may be allocated responsibility for initially determining the trustworthiness of any of its nationals employed in key positions by the multinational enterprise. But will the enterprise be required to accept as dispositive the nominating government's clearance? And will the host government be similarly bound?

On the other hand, the multinational enterprise may be given full authority to hire key employees on the basis solely of expertise. If so, the enterprise may hire persons viewed as security risks by one or more participants, including perhaps the host government. Given the manifold questions and their delicate character, it is clear that this area should be carefully considered in advance of launching a multinational enterprise.

Access limitations and inspections seem easier to handle than security clearances in a multinational setting. However, arrangements in this respect should be carefully spelled out in advance. If the host government also has primary operational responsibility for the enterprise, other participants being relatively passive investors, then the host government may simply specify and implement whatever access limitations it believes to be appropriate. If operational responsibility is multinationalized, then there are two general alternatives: (1) the access limitations may be mutually agreed and subsequently administered by the enterprise; or (2) they may be mutually agreed and subsequently administered by host government security personnel stationed at the facility.

The options regarding work-area surveillance are similar to those concerned with access limitations. In addition, it is arguable that if the enterprise is multinational at the operating level, employees of different nationalities may have greater incentives to monitor each other's behavior than if all employees are from one country. Whereas employees of diverse nationalities may reduce the risk of governmentally authorized diversion of nuclear materials for use in a weapons program, it seems doubtful that the presence of diverse nationalities will significantly affect the risk of employee theft or sabotage.

Physical containment and security forces. As discussed previously, a variety of physical containment measures—supplemented by a security force or combination of forces—are the main means for protecting sensitive nuclear materials and equipment against external threats of theft and sabotage. Physical containment measures are passive. They are an integral part of the facility within which a multinational nuclear operation is to be conducted. Agreement on the nature and extent of such measures, or a procedure for reaching agreement, should therefore be part of the basic agreement among participants covering the multinational nuclear enterprise. The containment measures will be largely implemented in the course of constructing the facilities involved. So all participants, and the host government especially, will have an interest in monitoring the installed systems to assure that they continue to operate effectively.

Physical containment should not raise problems for a multinational nuclear enterprise that are more difficult to resolve satisfactorily than numerous health and safety features built into the facilities involved. One issue that should be thought through in advance, however, is a procedure for back-fitting to strengthen containment measures in the event the assessed threat of theft or sabotage increases in the host country. Such back-fitting, like other physical alterations after construction has been completed, may be very costly. The host government and the other participants may, however, not agree that the external threat has increased sufficiently to warrant the addition of expensive new barriers or other containment measures.

As discussed previously, the host government will surely retain primary responsibility for any off-site security forces that are available in case of a threat of theft or sabotage at a multinational nuclear enterprise. Who would have authority over and control of an on-site security force at a multinational facility? The answer depends largely on whether the governments involved conclude that the threat is serious enough and the reaction of off-site forces is dubious enough to warrant an on-site security force that is capable itself of defeating a wide range of theft and sabotage attempts. The stronger the on-site security force, the more likely that responsibility for providing the force will be allocated to the host government. Indeed, the host government may resist the arming of any on-site security personnel who are internationally recruited. On the other hand, if the participants conclude that armed guards are necessary to protect a multinational nuclear enterprise, it seems doubtful that they would object to the guard force being supplied by the host government.

The security situation at IAEA headquarters in Vienna provides a useful example (although the analogy should not be pressed very far) of how participants and the host government might be expected to allocate authority to the host government in case of a heightened security risk. Following the 1975 Arab terrorist attack on OPEC headquarters (also in Vienna) and the kidnapping of several high officials of OPEC member countries, the Austrian government and the IAEA Secretariat quickly agreed that security should be increased at agency headquarters and during transport within Austria of laboratory samples of sensitive materials. In order to accomplish this result, armed Austrian police were provided both within agency headquarters and in escort vehicles accompanying shipments of material.*

Whether provided by the host government or the multinational enterprise, an on-site security force would need to be carefully coordinated with available off-site forces. Such coordination would be more difficult to arrange if the force itself were multinational than if it were provided by the host government. As mentioned above, it seems unlikely that the participants in a multinational nuclear enterprise would find an IAEA security force to be especially effective or politically acceptable.

Finally, it is worthwhile considering how the incentives for theft or sabotage by a criminal group would be affected by multinationalization. Assuming physical security systems are equivalent, whether a criminal group would prefer to attack a national or a multinational facility in a particular host country would depend largely on the group's motives. An attack on a multinational enterprise would involve more countries than the host government. This may or may not be advantageous from the group's viewpoint.

Industry Structure

The exploitation of economies of scale in fuel-cycle operations and the co-location of fuel-cycle facilities, especially reprocessing and fuel fabrication, would have advantages from a physical security viewpoint. Plutonium in a form

*The information on the response to the attack on OPEC headquarters was given to the author in discussions with IAEA staff.

that could be used directly in crude nuclear explosives would be located in relatively few central locations where special attention could be given to security measures. Moreover, co-location of reprocessing and mixed oxide fuel fabrication would eliminate the transportation link that is generally considered to be the most vulnerable to nuclear theft.

Multinationalization of fuel-cycle operations would enable the participants to realize these physical security gains in circumstances where pursuit of independent national programs would otherwise result in the dispersal of plutonium in many more locations and the multiplication of transportation links between facilities. In order to accomplish this, however, it is necessary to think through and plan the entire back end of the fuel cycle—reprocessing, mixed oxide fuel fabrication, and radioactive waste disposal. The interaction among these operations will be complex indeed, and an uncoordinated approach is not likely to produce the desired results from a physical security standpoint or, for that matter, from an economic viewpoint. On the other hand, rationalization of the entire back end of the fuel cycle may be too large a political bite for any multinational scheme to digest.

CONCLUSION

Whether or not multinationalization will reduce or increase risks of theft or sabotage will depend on the form and location of the enterprise, the precise nature of the physical security arrangements applied to the enterprise, and—most importantly—the future course of crime and terrorism in the nuclear age.

If nuclear operations are multinationalized, the host government is likely to insist on physical security measures that are at least as stringent as those for a national or private enterprise subject to its jurisdiction. At the same time, the other participants will want to be sure the host government, as well as criminal groups, do not steal nuclear material from the facility.

If designed to be reasonably effective, the physical security arrangements at a multinational nuclear enterprise seem likely to reduce the risk that any participating government will seek to divert material from the facility for use in a nuclear weapons program. Hence, multinationalization and physical security will both contribute to reducing the risks of nuclear weapons proliferation to additional governments.

If economic considerations dominate the timing, scale and location of fuel-cycle facilities, the worldwide nuclear power industry is likely to develop along lines where the problems of physical security will be manageable. If, however, nuclear nationalism prevails, and numerous small-scale facilities become widely dispersed, the problem of security against theft and sabotage may prove to be unmanageable. It is ironic, although true, that in attempting to strengthen its security by pursuing self-sufficiency in nuclear power, a nation may be reducing its internal security against criminal terrorists.

REFERENCES

U.S. Atomic Energy Commission, *Draft Generic Environmental Statement Mixed Oxide Fuel* (WASH. 1327, August 1974), V-12.

USNRC, Press Release No. 75-270 (Nov. 12, 1975).

Willrich, Mason, and Taylor, Theodore B. *Nuclear Theft: Risks and Safeguards.* Cambridge, Mass.: Ballinger Publishing Co., 1974.

Part III

Institutional Arrangements

Chapter 11

Institutional Arrangements for a Multinational Reprocessing Plant

Constance B. Smith and Abram Chayes

Institutional Requirements

The preceding papers have examined a number of the functional and technical aspects of reprocessing and recycle from the perspective of multinational operation: economics, the problems of safeguarding, health and safety, waste management, and physical security. From these analyses it emerges with reasonable clarity that multinational operation need not necessarily entail a penalty on any of these dimensions.

Indeed, in some respects the multinational approach holds out the possibility of substantial gains. The most important of these are the potential economies of scale that could be achieved in a large plant, serving, say, 50 reactors from a group of countries, as compared with a number of smaller national plants in those same countries, each geared to its smaller national clientele. Similarly, multinational operation would appear to offer opportunities for more effective safeguarding. Participation of the International Atomic Energy Agency (IAEA) in the design stage and maintenance of continuous on-site inspection might well be facilitated because the issue of national sovereignty need not intrude so sharply into relations between the agency and the operator. Governmental diversions would obviously be more difficult to arrange in a facility in which a single country did not have control of the entire operation. Although perhaps too much should not be made of it, the participants in such an organization could, in some degree, help keep tabs on one another. Again, with physical security, the multinational approach may enhance the possibilities for "rationalizing" the back end of the fuel cycle–for example, by co-location of reprocessing, fabrication, and storage facilities–and the joint concern of the parties to prevent theft or sabotage may prove advantageous.

Whether these gains will be realized–and indeed whether it will be possible to operate without penalty in the respects discussed—will depend heavily on the institutional structure that is adopted. Moreover, these institutional arrangements will have to meet additional requirements not dictated directly by the economic and technical considerations already examined.

Perhaps the most important of these is *assurance of supply* of reprocessed fuel or its equivalent as needed for reactor operations in a participant state. The OPEC experience has driven home the dangers of relying on others for vital energy resources. A principal rationale for acquiring a national reprocessing capability is to escape energy dependence. It is true that reprocessed fuel can provide only 20-30 percent of the requirements for a light-water reactor (LWR), and the remainder must be obtained elsewhere, in any event. Nevertheless, reprocessing could be a significant source of supply, and the institutional arrangements must provide a high degree of assurance that it will be available when and as needed for reactor operations. This means not only that there should be no express political strings attached to the participant's entitlement; the decision processes must also be such as to preclude the introduction of political considerations *sub rosa.*

A related problem is the disposition of the plutonium produced from reprocessing. All agree that it would be folly to return the plutonium in pure form to the state from which the spent fuel came. To do so would be to provide it with direct access to weapons material (and to increase significantly the risks of theft). Plutonium mixed with uranium or even fabricated into fuel rods is not much better. The alternatives are to send the plutonium to a nuclear weapons (or breeder) state in return for an equivalent amount of enriched uranium, or to fabricate it into mixed-oxide fuel rods to be stored in the multinational center until actually needed for reactor loading. Suitable arrangements for exchange, storage, and release must be constructed without compromising security of supply.

A second major requirement is *political acceptability* to states that might otherwise build national reprocessing capacity. Advanced developing countries, which are particular objects of concern, are also particularly sensitive to the fact or appearance of discrimination in arrangements that include both developed and developing countries. And this sensitivy seems to be most acute in matters relating to technology transfer in general and nuclear energy in particular. The arrangements must, at a minimum, provide for meaningful participation in the operation by developing country parties. Every effort must be made to minimize any differential unfavorable to them—the closer the approach to an equal basis of participation for all, the better.

Again a related problem affects the nonproliferation impact of multinationalization—the problem of technology transfer. There is some dispute as to the current availability of reprocessing technology. The basic information has been in the public domain since 1958. It seems clear that a considerable number of states, including some developing countries, have the capability of translating this information into a small reprocessing plant to handle low-burnup fuel from a research (or production) reactor.

At least a dozen such plants are now in existence, of which two or three are in developing countries. Large-scale commercial reprocessing of high-burnup power-reactor fuel is another matter. The difficulties are sufficiently illustrated by the experience in the advanced countries, none of which has yet been able to keep a reprocessing plant for LWR fuel in operation on a commercial basis for any considerable period. The commercial technology is hard-won and expensive and it is controlled by its possessors as a valuable item of industrial property.

Access to this technology—and especially to know-how and operating and fuel-handling experience—by a participant in a multinational facility would be useful not only in the construction of a national commercial plant, but also in constructing and operating a small-scale plant to produce weapons material. So if all the participants in a multinational enterprise are given access to design and operational details, there is some danger of spreading technology and know-how that is useful for weapons production The exact magnitude of this danger is unknown and depends on disputed judgments about the current availability of the relevant technology and capability. We have made the conservative assumption that this is a real problem, and that significant attention should be given in the institutional arrangements to minimizing the diffusion of technology and know-how. This assumption, however, cuts across the objective of nondiscrimination.

From the foregoing discussion we may derive a set of basic institutional objectives or desiderata that will decisively shape the organizational decisions. What are these minimum institutional objectives? Or to put it somewhat differently, what must the multinational enterprise do for nonproliferation and what must it do for the participants?

To satisfy its nonproliferation objectives, the organization must discourage the building of national reprocessing plants by providing a viable alternative to a national reprocessing facility. If it does not dissuade countries from building national plants—or worse yet, if it would facilitate the building of such plants by spreading reprocessing technology and know-how—then it would have failed as a nonproliferation measure.

While discouraging national plants, the multinational activity must not aggravate proliferation concerns in other ways. It must, for example, do everything possible to minimize the risks of theft or safeguards evasions.

To be an attractive alternative for potential participants, the facility must provide an assured source of supply. There can be no politically motivated terminations or suspensions of services. It must also guarantee economically efficient and reliable services at a cost less than or equal to that of performing these services nationally. Finally, it must provide the consumers with a level of participation that is sufficient to counter domestic pressures against participation in a multinational enterprise controlled at least in part by the developed countries.

These objectives are by no means wholly complementary or even compatible. We have already noted that limitations on technology transfer may be inconsistent with the goal of eliminating discrimination among the parties. Indeed, to the extent that an incentive for participation is to gain experience with the technology, that incentive will be seriously weakened by stringent constraints on access to technology. Similarly, the need to control the reprocessed plutonium may in some measure cut across the requirement for security of supply. And if we seek to deal with that problem by rigid provisions in the organizational charter to establish the entitlement, we may compromise the flexibility required for effective commercial operation.

The overarching tension or trade-off is that between the commercial and political aspects of the enterprise. From one perspective, a multinational fuel-cycle center is designed to provide an industrial service in a technically sound and economically efficient manner. Operational decisions must be responsive to the requirements of the technology and the dictates of the market. But from its beginnings, the nuclear industry has been freighted with political implications. It is so regarded and so handled by all governments. Moreover, a multinational center would itself be designed to achieve certain political objectives. A pervasive concern in the design of the institutional arrangements is to reach a proper balance between these two aspects, both so that decisions that are properly commercial are made by technical people seeking to rationalize the operation of the enterprise in technical and economic terms, and that questions that are properly political are consigned to political organs manned by government representatives expressing governmental interests.

The Elements of the Organizational Model Employed

The rubric "international arrangements for the nuclear fuel cycle" could properly be applied to a wide range of activities and organizations. At one extreme the term could comprehend, for example, a loose set of arrangements designed to ensure the participants' supplies of reactor fuel, but without any formal organizational or institutional embodiment. Or coming closer to what has been the focus of our discussion, it could cover a variety of forms of international participation in enrichment and reprocessing activities. A subsequent paper [Chapter 12], for example, describes the setup of Urenco/Centec and United Reprocessors, which take the form principally of entities for exchange of information and coordination of three national programs, at both the industrial and political levels. In Eurodiff, a number of countries and enterprises have made capital investments, carrying some right to a portion of the output, but no participation in management and operations. Eurochemic provided a more intimate involvement of relatively homogeneous countries in the operation of a chemical reprocessing plant. No one particular form or organizational structure can claim a monopoly on the title of this symposium or even a right to be regarded as an ideal type.

For the purposes of the present discussion, we have envisioned a rather complex international organization, established to own, manage, and operate a large-scale chemical reprocessing plant and related facilities. And we have tried to develop a rather detailed analysis of the institutional problems involved in putting such an organization in place. We do not regard this model as a blueprint to be translated into reality by any existing group of countries. Obviously any practically functioning organization would have to reflect the particular requirements and objectives of its members. Our model is designed to expose for analysis the problems and issues of institutional design of any such organization, and in particular to identify the likely political pressure points in the negotiation. For these purposes, our model is more complicated and taxing than any real-life multinational enterprise is likely to be. This intensifies the institutional problems, but if we can see the direction of likely solutions even within the framework of our demanding assumptions, we can perhaps be hopeful that solutions would be found in more realistic situations.

In the model for analysis, we have made the following assumptions:

A high degree of governmental involvement in the enterprise. It is conceivable that a number of operating entities might join together to build and operate a reprocessing facility, just like any other transnational investment or enterprise. The rights and obligations of the parties, their required contributions, and their participation in management and control would be determined by agreement between the operating entities that were the participants, and there would be no special intergovernmental involvement or agreement. Each participating entity would, of course, be subject to the laws of its own country, and the joint enterprise would have to comply with the regulations of the host country. In addition, if a participating entity were publicly owned or otherwise subject to governmental policy direction—which means in almost every case but the United States—the home government could control the entity's actions within the joint enterprise. But there would be no direct or ongoing government participation in the enterprise.

In practice, it is hard to conceive of any arrangement going to this extreme of laissez faire. Even today, reactor and other equipment sales that are treated as commercial transactions are invariably carried out under the umbrella of an intergovernmental agreement specifying among other things the undertakings of the government of the purchaser with respect to protecting against proliferation and other hazards. It seems unlikely that governments would wish to go below this minimum in connection with the development of a multinational reprocessing facility.

It is possible that in some fairly simple cases the governments would be satisfied with little more. One can imagine a few contiguous and culturally homogeneous countries banding together to buy a reprocessing plant to service their joint needs in a manner not much different from other international utility enterprises. The supplier entity might be a participant in the venture, or simply a contractor to build and operate the enterprise.

But situations of this kind are likely to be rare, given the high political visibility of almost every aspect of nuclear energy, and especially the parts involving potential weapons materials. It seems likely therefore that governments will not content themselves with an advance agreement laying down the terms and conditions within which the enterprise would function, but will insist on some kind of continuing control.

It is clear, however, that even if the enterprise is to be heavily dominated by governments, an intergovernmental organization on the lines of the typical U.N. specialized agency would be wholly inappropriate. Those organizations are designed to perform essentially legislative and regulatory tasks—and they are not too successful at that. They are structurally unsuited for operational and managerial functions. So we assume that the organization would be cast in some version of what we may call the "corporate" form. By this is not meant that the venture would necessarily be incorporated under the laws of one of the participating countries or a special statute—although that is not excluded. Eurochemic took this route, as do many government-owned utilities and other enterprises. Intelsat did not, although the Definitive Arrangements recite that it shall have "juridical personality." The essence of this form, in our view, is to be found in two principal features: (1) a separation of ownership and control, which is necessary if governments, at the political level, are to be kept out of day-to-day management, at least to a reasonable extent; and (2) a pool of capital subject to the control of the enterprise. Whether raised by subscription from the members, the market, or otherwise is not crucial, but the enterprise should not be dependent on periodic contributions from participating governments—otherwise these governments, especially the major ones, would be able to impose their will by withholding contributions.

It might be thought that limitation of liability would be another essential of the corporate form. But no one—especially not the host country—will be willing to accept an arrangement in which liability, at least for nuclear accidents, would be limited to the assets of the enterprise. It is probable that there would be special arrangements covering compensation for damages from a nuclear accident, as is the case with other nuclear activities.

The participation of a substantial and diverse group of countries. Our model does not specify an exact number of countries, but we are thinking in terms of a dozen or more. Moreover, we stipulate that the membership would include both developed and developing countries, that some of the participants would themselves be suppliers of reprocessing equipment and technology, and that some, probably the large majority, would be only consumers of reprocessing and fuel-cycle services. On the record so far, these assumptions are to some extent contrary to fact. With the exception of Eurochemic, in which thirteen countries participated, all of the multinational fuel-cycle ventures to date have involved only a handful of countries—less than half a dozen. Again, except for the Iranian connection with Eurodiff (which is limited to capital investment), all the participants have been advanced industrial countries. Groupings that form the basis of current speculation or study are also, in general, made up of only a few nations, usually with some regional or other extrinsic basis of common interest and concern. Here is an instance, therefore, where our assumptions are much more exigent than those likely to be in fact encountered. Nevertheless, we think they are useful to highlight aspects of the political dynamic that will inevitably be involved.

The organization will be empowered from the beginning to provide the full range of fuel-cycle services. These would include spent-fuel storage, reprocessing, fuel fabrication, and storage of plutonium (whether in the form of mixed-oxide fuel rods or otherwise) pending authorized release. We recognize that, for a variety of reasons, the organization will probably not wish to begin all these functions immediately, and this introduces a new set of questions about phasing and staging of the various activities. But we assume that authority for all of them will be included in the charter from the outset. This assumption, too, may be more demanding than necessary. It is possible, perhaps likely, that any multinational venture would begin on a more limited basis—with spent-fuel storage or the establishment of a plutonium depositary. The question whether to proceed to more complex activities could be left for future determination. In that case the institutional arrangements could be considerably less elaborate.

ELEMENTS OF THE ORGANIZATION

Constitutent Instruments

The formal structure of a multinational fuel-cycle facility would be embodied in one or more constituent instruments negotiated among the parties. Certain fundamental decisions will have been agreed at the outset and will be incorporated in these constitutive documents. If, as is likely, the key provisions of the constituent instruments can only be amended by a lengthy and difficult process, then providing for any particular matter in the constituent agreements will have the obvious advantage of certainty. All parties will know at the outset how that question is resolved, which relieves them of the risks of a later ad hoc decision. This certainty does, however, have a cost—rigidity. The use of pre-set rules eliminates the flexibility that is possible when decisions are made on a case-by-case basis as they arise. The need for certainty may override the desire for flexibility, but relative costs and benefits of certainty and flexibility will vary depending on the type of decision in question.

Among the matters that will almost certainly be established in the constituent instruments are the following:

- The identity of the initial participants; the qualifications of additional participants and the procedures for adherence; withdrawal or expulsion, including any right to a share of the assets in such case.
- The decision-making organs, and their powers and procedures.
- The rights and obligations of the parties.
- Initial capital contribution, and obligations with respect to additional capital.
- The general location of the facility.

A number of major questions will not be susceptible of decision in advance, and as to these, two approaches—not necessarily mutually exclusive—may be adopted. The parties may seek to establish, in the constitutent instruments, substantive norms or guidelines to govern future decisions. And they may assign particular kinds of decisions to different organs or decision procedures deemed likely to produce acceptable results. In this regard, the key problem, as already noted, will be to distinguish between questions that should be decided by governments in their political capacity and those to be decided at the operational level and insulated as much as possible from political considerations. Considerable insulation can be achieved even if a country participates in the enterprise through an operating government department or a government-owned company. Although such organizations do not have the autonomy of private corporations, they often have considerable independence from political control and tend to act much like their private counterparts in business matters.

It is sometimes useful to segregate the two categories of question—political and operations—in two separate agreements. Both Intelsat and Eurochemic followed this course. Matters with high political content or engaging the obligations of governments as such were dealt with in an agreement among governments. Operational matters were dealt with in a separate instrument to which the operating entities were parties. If the organization has the form of joint venture, this instrument would simply be a contractual agreement among the parties. Another possibility would be to create a separate juridical entity by incorporating the enterprise under a special or general statute.

Decision-Making Bodies and Procedures

Assuming, as proposed above, that the organization is cast in the "corporate" form, the principal decision-making organs would most likely be: (1) officers in charge of day-to-day management (hereinafter called the management) headed by a general manager; (2) a policy-making committee (hereinafter called the governing board); and (3) a general organ in which all investors in the enterprise would participate (hereinafter, the Meeting). The Meeting might be dispensed with if the group of investors were small enough to allow each investor to be represented on the governing board. Nor is it excluded that the supplier entity would be appointed general manager, like Comsat in the early Intelsat arrangements.

In addition to these three groups, there might also be an external control organ composed of government representatives (hereinafter, the Assembly). The decision whether to have an Assembly will depend upon the degree of control that governments wish to exercise over the enterprise. While reprocessing is not now generally considered to be a vital service, it could become so in the future if the price of uranium were to continue to increase or if breeders were to become widespread. If governments perceived reprocessing as a vital utility, they would become increasingly reluctant to allow unfettered operation of even domestic reprocessing facilities. They would be even more hesitant to rely on an international reprocessing group over which they could exert no control. Sundström (1972: 14–15) describes the problem very succinctly:

> Nor is this desire for political control hard to understand. For example, if even a significant part of the electricity sources were controlled by an external agency not supervised by the state, and that agency should decide to terminate deliveries for some reason or another, or even only to raise rates significantly, the recipient state would find itself in unacceptable political economic dependency. Even the theoretical possibility of such an eventuality makes governments consider very carefully with whom and under what conditions they enter into a joint enterprise.

It is clear, therefore, that the governments will want to exert some degree of control over the commercial entities involved in the enterprise. This control could be exerted in several ways.

One mode is the use of direct controls by each government over its participating entity. These could include licensing and policy controls, directives to the entities involved, and a requirement of regular reports and briefings by them on the operation of the plant.

One problem with trying to influence the nature and operations of multinational facilities through pressure on the participating industries is that this may dilute their ability to act as commercial partners rather than government representatives. This problem can be partially alleviated in the government-level agreement, which would deal with major issues of importance to governments. But there is much to be said for establishing a mechanism by which the governments can meet and develop common policies directly rather than indirectly through government pressures on the operating entities.

If more direct government control is desired, it could be accomplished by building control measures into the corporate structure or through the use of a special government control organ. Types of control measures built into the corporate structure would include government representation in the Meeting and governing board, government power to appoint board members and officers, a special requirement of approval by governments for certain decisions, and government fiscal control and auditing powers.

If a government control organ were established it could be given varying degrees of power. The Intelsat Assembly of Parties meets only every two years, for the most part to consider matters of general policy and to make recommendations to the other Intelsat bodies. But with a multinational fuel-cycle enterprise, it is likely that an Assembly would meet more frequently and have power to bind the enterprise in some respects. The government control organ need not be manned by foreign-office representatives, as it is in Intelsat's Assembly of Parties. Eurochemic, for example, uses a Special Group of the Steering Committee of the European Nuclear Energy Agency as an external control organ. All participating governments are represented on the Special Group.

The Special Group of Eurochemic governments is an interesting example of an external organ given substantial powers. Under the Convention, the Eurochemic Company must submit a yearly report on its development and financial position to the Special Group. The Special Group, on its initiative, may consider any problems of common interest to the governments raised by the operations of the company and may make recommendations. In addition, any director or shareholder of the company may submit to the Special Group difficulties arising in connection with processing of fuel, allocation of products, the use of resources of the company for research and the communication of the results of research. Decisions on these subjects adopted by a three-fourths majority of the members of the Special Group are binding on the company. Certain amendments of the Statute of the Company on such matters as its objects, the conditions for admission of new shareholders, the adoption of decisions by the General Assembly and the composition and adoption of decisions by the Board of Directors, require unanimous approval of the Special Group. Other statute amendments, as well as increases or reductions of capital which would result in changing the distribution of capital among the shareholders, require a decision adopted by three-fourths of the Special Group. (See Fligler, 1967: Annex, for the text of the Eurochemic Convention and Statute.)

The use of an external control organ similar to the Eurochemic Special Group has the important advantage of preserving government control over certain crucial matters, such as some amendments of the constituent instruments and authorization of capital, while allowing latitude for management independence on operational questions. Independent management is important for several reasons. First, it is more efficient and flexible, which will make the goal of maximum economic efficiency easier to achieve. Of even greater significance, however, is the importance of an independent, nonpolitical management for security of supply. If decisions about such things as scheduling and queuing are politically motivated or controlled, then a political (or apparently political) slowdown or even interruption of reprocessing is much more likely than it would be if those decisions were made by industry types, motivated by economic concerns. Thus, paradoxically, governments might be better able to assure security of supply not by giving themselves control over management, but by giving nonpolitical industry that control. In other words, "to reap the benefits of joint operations while at the same time protecting the corporation from pressure from the other states, each state [may be] willing to give the corporation substantial autonomy where such relinquishment of influence is multilateral" (Sundström, 1972: 10).

An extreme example of the ability of a politically independent management and board to carry on the operations of a joint enterprise once established, despite subsequent political disagreements between governments, is seen in the case of the Central African Power Corporation. The corporation was established by the British when they dissolved the Federation of Rhodesia and Nyasaland in 1963 to operate the large hydroelectric facility at the Kariba Dam on the border of what became Zambia and Rhodesia. Although Zambia refused to acknowledge the white minority Rhodesian government and, consequently, the government control organ has never met, the hydroelectric facility continued to operate under the management of the board members appointed before the Rhodesia-Zambia conflict. This type of arrangement is far from ideal or stable, but it does show that a multinational facility supplying a vital service, here electricity, can continue to operate for at least a time in periods of complete government-level conflict, if the management does not require political agreement or direct government participation. (See Sundström, 1972: 63, for a full discussion of this enterprise.)

Because of these advantages of nonpolitical management together with the possibilities of creating an external government control organ to protect major governmental and political interests, it is likely that a multinational fuel center should have a government organ. Similarly, the need for economic efficiency and security of supply dictates the appointment of a nonpolitical management responsible to no single government. Members of the management and staff might be governed by the traditional provision applicable to international civil servants forbidding them to "seek or receive instructions from any Member or from any other authority external to the Organization."

Unlike the management, which would be constantly involved in the operations of the plant, the Meeting and the governing board are likely to meet much less frequently and thus exert less direct influence over operations. The management would be selected by the governing board and responsible to it, although the general manager might be subject to approval by the Meeting. Management autonomy from the governing board would be further strengthened if officers could be removed only for cause by a majority of the board. Beyond that, the board's primary responsibility would be to oversee the actions of the management and to set general policies within the framework of the constituent agreement.

If norms for the most crucial questions, such as pricing, scheduling, and queuing were established in the constituent instruments, the governing board would not be able to interfere seriously with security of supply. (The problems involved in setting such policies in the constituent agreements will be considered below.) If the governing board's discretionary authority is thus limited, then it could tolerate some government influence, although such influence would not be particularly desirable in light of the close working relationship between the board and the management, which must be nonpolitical. On balance, therefore, an apolitical governing board would be preferable. W. Friedmann, for one, has argued that all personnel in an international public corporation, from the board of directors downward, must be selected and hold their posts on a truly international basis, detached from national loyalties, quotas, and policies (Friedmann, 1943: 197). If basic policy and crucial political decisions are made by the governmental control organ on which all governments are represented and given an equal vote, then the need for national representation on the governing board is greatly reduced, at least as long as no one member is given a majority on the board.

Various arrangements for selecting the governing board are possible. Members might be elected by the Meeting, representing the investors who put up the capital for the project. This would be most natural if the enterprise were organized as a corporation with shares representing investment. But even in a joint venture, voting for directors could be based on the amount of quota paid by each investor. Such an arrangement would have the advantage of giving greater voting power to those with a greater financial stake in the enterprise. Minority representation could be achieved by a provision for cumulative voting.

If, as is likely, the number of participating states is small, then each country might want a representative on the governing board. This might tend to turn the board into a group of national representatives rather than an international body. In any case, the impact of political considerations on the governing board can be minimized if it is understood that members will have a technical and operational—rather than political—background.

Voting procedure in the decision organs is characteristically an area of intense negotiation in the establishment of international institutions. In principle, as we have suggested, the basic rule would probably be one-country-one-vote in the Assembly and voting in proportion to investment (or some quota formula) in the Meeting. In the traditional corporation, each governing board member has one vote. But in many international organizations, including some of the most successful (such as IMF and Intelsat), the vote of each board member is weighted in accordance with the voting power of the participants that elected the member.

These basic rules will be overlaid with a variety of special provisions for particularly important questions: special majorities, the requirement of concurrence of two or more organs, procedures for breaking deadlocks.

A question of some general importance is whether, in an enterprise with both supplier and consumer participants, the concurrence of each group, voting separately, should be required for some decisions. United Nations rules for commodity agreements require such a distributed majority for major decisions, and the procedure would have special relevance in our case for decisions with heavy implications for proliferation. On the other hand, division into supplier and consumer groups may be seen as stressing or even entrenching an unacceptable discriminatory relationship. This issue will be resolved by negotiation in each case, but in the materials that follow we suggest some instances where supplier concurrence would be desirable from a nonproliferation standpoint. If the consumers will not agree—or in an all-consumer enterprise—a possible, if unorthodox, alternative would be to provide for an IAEA veto on such issues.

Finally, there is the problem of the settlement of disputes that may arise in the course of operating the enterprise. Constituent instruments of international institutions often provide for arbitration of disputes of a legal nature or involving interpretation of the basic agreement. Arbitral procedures might have a considerable role in relation to essentially commercial disputes, and perhaps in connection with disputes over a consumer's right to reprocessed fuel. But on issues with strong political overtones, arbitration has not proved a satisfactory mode of dispute resolution. In a number of international organizations—the IMF and some commodity agreements, for example—disputed questions of interpretation are remitted for ultimate resolution by a special majority of the highest decision-making organ (here, the Meeting or the Assembly), sometimes after an advisory opinion from a legal panel. This process may derogate from the strictly legal value of the arrangements embodied in the agreement, but it has the advantage of permitting adaptation of the enterprise to changing circumstances and needs of the participants.

Participants

As already noted, we have assumed that the organization will include one or more countries that could provide reprocessing technology and equipment and a larger number of others that would be exclusively users of the services to be provided by the center. (Although the terms are not exact, we have referred to such countries as suppliers and consumers, respectively).

Suppliers. A country that supplied all or part of the equipment, technology, and know-how might well provide long-term *financing* for the plant at favorable repayment terms. In addition, other supplier countries interested might also agree to give direct subsidies to the plant, in view of its potential nonproliferation impact. Such subsidies would certainly make the plant more economically attractive, but they must not be perceived by consumers as an effort by the supplier to buy them off.

In any case, supplier participation will be crucial at the *design and construction* stage. In particular, one or several suppliers could be expected to design the facility, provide the necessary equipment, and assist in the construction, start-up, and training associated with the facility. Having the supplier build the facility has the side benefit of limiting the spread of reprocessing technology. It would also be politically acceptable to the consumers if they did not already have the necessary technology and equipment but wanted a reprocessing facility.

A turnkey operation, with the termination of supplier involvement ending at the construction stage, could lead to a multinational structure that is not desirable from a nonproliferation perspective. A preferable arrangement would be one in which the supplier also participated in *business policy-making,* in the *management* and in the *technical operation* of the plant. However, as will be noted below, significant and continuous supplier involvement and control may not be politically acceptable to some potential suppliers and recipients. As a result it might be necessary to have active supplier involvement in the early stages of the operation of the facility, and then gradually phase out the supplier role in the later stages. This reduction in supplier participation might be accompanied by an increase in IAEA involvement in the operation of the facility. IAEA operators would certainly be suitable monitors, and, at the same time, they might be more politically acceptable because they represent an international organization, not a single supplier country.

Membership on the governing board would give the supplier a voice in the general policy and operations of the enterprise, including the nature and frequency of IAEA inspections, physical security measures, and the general conditions for supply to members and third parties. Supplier participation on the board may also be useful in giving the supplier a more direct stake in preventing any diversions from the plant. As indicated by Canada's experience with India, no supplier wants to be implicated in the creation of a new nuclear weapon state. Being a board member, as well as a supplier, of a facility that allowed sufficient diversion of plutonium for the building of a nuclear weapon would be an especially embarrassing position.

Voting arrangements could be worked out, if necessary, to make the arrangement politically acceptable to the host country and other participants. Moreover, this form of participation would not be particularly onerous for the supplier entity. Nevertheless, participation limited to membership on the governing board is unlikely to afford significant control over the actual operations of the plant. As a result, this form of supplier participation by itself may not be adequate either for efficiency of operation or for nonproliferation purposes.

From a nonproliferation standpoint, having supplier representatives in the management and on the technical staff would be the best form of supplier involvement. It would give the supplier influence over day-to-day operations and would provide opportunities for continuous observation of the activities of other operators and managers. Officers and employees from the supplier country would be assigned specific operational tasks but, in addition, they would, like other participants, remain alert to possible diversions. Since the supplier presumably already has the reprocessing technology, there would be no technology-transfer problems if the supplier helped run the plant.

Since participation by supplier personnel at the management and technical levels will probably be necessary, at least in the initial stages, for the efficient operation of the plant, their presence may not meet with any serious objections. On the other hand, consumer countries may want to limit the number and duration of supplier personnel or functions performed by them. Even if the consumer countries agree to having supplier involvement in management and operations, the supplier itself may be unwilling to undertake a long-term commitment of this nature. It may, however, be possible to limit the number of supplier representatives to a level that is sufficient for operational and monitoring purposes but also small enough to be no real burden on the supplier or irritant to the consumers.

Consumers. It is in defining possible types of consumer involvement in a multinational facility that the tension between promoting nonproliferation objectives and ensuring political acceptability becomes most dramatic. The best scenario from a nonproliferation standpoint might be a regional plant, in a supplier country, owned, built, and operated by a supplier and selling reprocessing services to a number of countries in the region. Such an arrangement would, however, probably not be an acceptable alternative to national plants. In particular, it would make consumers dependent as a practical matter upon the good will of the supplier and thus not give them the assurance of supply they desire. On the other hand, the most politically acceptable structure might be a multinational plant, owned and operated by a group of consumers with free access to reprocessing technology. This, of course, would be unacceptable for nonproliferation purposes because it would facilitate the building of national plants by giving the consumers the necessary reprocessing technology and operational experience. The ideal structure must lie somewhere between these extremes.

Consumer contributions to capital and similar participation in *financing* is a desirable form of involvement both economically and politically. On an economic level it would provide needed capital for a capital-intensive enterprise. On a political level it would increase the political acceptability of the plant by giving consumers an ownership share. On the other hand, for developing countries—particularly those that are not oil producers—this form of participation may not be feasible. Large capital contributions, therefore, should probably not be a condition of membership.

Consumer participation in the *design and construction* of the plant would carry severe drawbacks for nonproliferation purposes. The technology-transfer problem is most acute at this stage because of the sensitive technologies and know-how involved. However, supplier design and construction should not cause severe political-acceptability problems as evidenced by the success of turnkey arrangements for reactor sales.

Having consumers on the *governing board* should cause no substantial proliferation problems and would contribute to political acceptability by giving participants a voice in the major policy decisions. One form of participation is to give representation to key investors, who have contributed a specified percentage of the required capital to be assured a percentage of the output, with voting rights related to the share of investment. But representation on the board need not be directly correlated to investment or use.

Consumer participation in plant *management,* like participation in design and construction, raises problems of technology transfer. The managers will be involved in the day-to-day running and maintenance of the plant, and will have authority to request information from the technical operators. These are arguments for limiting consumer involvement in management, to the extent politically possible.

Consumer participation in the *technical operation* of the plant creates even greater risks of technology and know-how transfer. The best arrangement for nonproliferation purposes would be a plant run entirely by supplier personnel. It is not clear, however, whether the consumer participants would agree to such an arrangement. If the supplier did not provide all of the operators, some combination of supplier and consumer operators would be necessary. The host country would have a strong claim to a role in the operations because of its special interest in health, safety, and security matters. Beyond this, however, all the consumer participants would seem to stand on an equal footing, and any attempt to disqualify nationals of a particular country from employment on the technical staff is likely to cause severe political complications. Although, as noted, this will increase the potential for technology transfer, there is reason to think that a broadly multinational work force would have some advantages with respect to diversion and security. It may be possible to minimize the technology-transfer problem by assigning employees from nonnuclear-capable countries to less sensitive areas of operation.

The technology-transfer problem would also be reduced by promoting loyalty to the enterprise among the technical operators. Employees should be paid well and given sufficient fringe benefits and respect to make them less likely to accept an offer to take their skills and work elsewhere. Their employment must be long-term, both to increase plant loyalty and to reduce the number of persons exposed to sensitive technology and knowhow. The kind of staff rotation permitted in Eurochemic should be avoided in a multinational facility seeking to minimize technology-transfer risks.

The host country. The foregoing comments will be equally applicable to the host country, in its capacity as either a supplier or consumer participant. But the country in which the facility is located would have special functions and roles by virtue of that fact alone. The most important of these relate to *health and safety* and *physical security*, although there would be a myriad of other problems in fitting the multinational enterprise into the general political and legal regime of the country.

Radiological emissions from the plant and the consequences of any accident will fall primarily on the host country. It seems logical therefore that it should have an important say in the design and operational program of the plant. If the country has an existing licensing or regulatory procedure, it might want the multinational facility to meet these domestic requirements. The special features of such a facility and the political sensitivity involved will probably make this course unattractive for both sides. It is likely that some special procedure—perhaps requiring specific legislative authorization—would be adopted. In any case, the host country is likely to insist on final approval of the health and safety features of design and operations. This has obvious implications for technology transfer. The host has a more plausible claim that it cannot perform the licensing function without full access to technical and operational details. These considerations should be kept in mind in the decision on the location of the plant.

The second major area of host country responsibility is the physical security of the plant and related facilities, including transportation routes within the country. It has sometimes been suggested that a multinational center might be able to establish a multinational security force to perform some of these functions. Perhaps this would be possible for the on-site guards. Even as to these, however, the host country will no doubt want a large say in the organization, makeup, training, and operating routines. For such a force would, in effect, be carrying out enforcement functions (probably armed) on the territory of the host country. Moreover, as in the health and safety area, the principal effects of a breach of security are likely to be in the host country.

As for off-site support forces, it hardly seems feasible that these could be organized on a multinational basis, given the sporadic and unpredictable occasions for their use. This support will almost certainly be provided by the law enforcement authorities of the host country (often under provincial or even local command and control), and ultimately the armed forces. The relationship between these forces, on the one hand, and the plant management and on-site guards on the other—including liason and communications, response routines, rules of engagement, and the like—need to be thought through and, to a large extent, negotiated in advance. The above sketch of the subject matter is enough to suggest the difficulty of the exercise.*

IAEA. It is possible to envision the IAEA playing a range of roles in a

*See Chapter 6, above.—Eds.

multinational endeavor. Political as well as non-proliferation and economic benefits could be visualized through IAEA involvement.

Beyond its safeguard functions, the IAEA could provide technical support in the *planning and design* of plutonium-related facilities, could provide some IAEA personnel to the *operating staff* of the plant, and might even be invited to become a member of the governing board and to take one of the *management jobs* in the plant in order to play a role as monitor. Traditional IAEA inspections would continue, but the constant presence of IAEA personnel as operators, for example, would serve to augment them. Furthermore, the presence of IAEA international civil servants on the governing board might serve to mute political wranglings over what should be commercial or technical matters.

Additional roles for the IAEA could involve the agency becoming a depository of excess plutonium over what is needed commercially, by managing a plutonium storage site associated with a multinational regional center, or, in the extreme, the IAEA could own the multinational plant. This latter possibility would involve a major shift requiring special financing and staffing for the agency, but arguably it would have the benefit of giving a true international character to the reprocessing facility. IAEA ownership might also help minimize the possibility of expropriation of a multinational facility if the plant is given the extraterritorial status granted such IAEA facilities as the headquarters in Vienna, Austria, and the Theoretical Physics Center in Trieste, Italy.

Although the IAEA has never undertaken anything of this scale, it has the authority under its Statute;

> to acquire or establish any facilities, plant and equipment useful in carrying out its authorized functions, whenever the facilities, plant and equipment otherwise available to it in the area concerned are inadequate or available only on terms it deems unsatisfactory (IAEA: Art. II, B.1).

On the other hand, as argued above, the facility could not be effectively operated as an international organization with the structure of a specialized agency and dependent upon yearly contributions from member states. Because of this and the IAEA lack of experience as owners and operators of major nuclear facilities, IAEA ownership seems neither particularly desirable nor feasible.

Article XII, A.5 of the IAEA Statute deals specifically with reprocessing and gives the agency the right, with respect to any agency project, or other arrangement where the agency is requested by the parties concerned to apply safeguards, "to approve the means to be used for the chemical processing of irradiated materials," and "to require deposit with the Agency of any excess of any special fissionable materials recovered or produced as a by-product" over what is needed for peaceful purposes. From the wording of this provision it is unclear whether the agency's right to require the deposit of excess special fissionable materials extends only to materials recovered or produced as a by-product of *reprocessing,* or whether it extends to *any* excess fissionable materials. The letter of transmittal from the Secretary of State to the President, sent in February 1957 (U.S. Senate, 1957) states that the deposit provision applies to any excess special fissionable materials from Agency-sponsored projects. It would, therefore, give the IAEA authority to require the deposit of excess spent fuel and reprocessed fuel from IAEA projects and IAEA-safeguarded reactors, since fuel would contain special fissionable materials.

The precise nature of IAEA involvement will ultimately depend on the confidence participating governments have in the agency, and on its ability to accept new responsibilities without compromising its traditional functions in operating the safeguards system and in technical assistance. These latter considerations probably argue against any significant involvement of the IAEA in operational activities.

Financing

The capital costs of large reprocessing plants are likely to be in excess of $100 million. The Allied-General Nuclear Services 1500-MTU/year plant being constructed in Barnwell County, South Carolina, was estimated in 1968 to cost $50 million (U.S. SAEC, 1968: 171). To date more than $250 million has been spent and an additional $500 to $650 million may have to be spent to build facilities to convert the plutonium nitrate and the high-level radioactive wastes from liquid into solid form (Tannenbaum, 1976: 1). A multinational plant might cost less than the Barnwell plant depending on design and safety equipment, but a large capital investment will, in any event, be necessary. Thus the question arises: How is this capital to be raised? Or to put it differently: What is to be the capital structure of the enterprise?

It would, in theory, be possible to finance a multinational reprocessing/ activity on a totally ad hoc basis. Instead of contributing equity capital to a corporation or similar entity, each participant might agree to pay an agreed share of the expenses as they arose. Such an arrangement would have several disadvantages. To begin with, the lack of paid-in capital would make it difficult if not impossible to obtain long-term loans, unless such loans were guaranteed by the governments. Creditors are much more likely to lend if they have a fixed amount of capital to look to in case of default. In addition, such an ad hoc arrangement would be especially hard to enforce as against government participants if they should default on their financial obligations. Perhaps more important, ad hoc financing would not give the enterprise that independent financial status that may well be necessary to insulate the facility from the undue political influence and uncertainty resulting from reliance on yearly contributions from member states.

Thus it would be desirable to ensure that funds sufficient to finance at least the initial investment program are safely in hand from the beginning. The initial capital might be raised in the ordinary capital market, but for a variety of reasons, a public issue does not seem appropriate for the kind of organization defined by our assumptions. The basic capital is likely to be raised by contributions from the participants–whether operating entities or governments–in an amount or under a formula specified in the constituent agreements. Such a contribution might take the form of a purchase of the shares of the enterprise, if it is a corporation, as was the case with Eurodiff. Otherwise it would simply be a payment into the capital account of the venture. Debt financing could be obtained from traditional lending sources as well as from governmental and international lending institutions.

The amount of capital to be contributed by each participant could be set in various ways. Participants might be left free to contribute what they chose or they might be required to contribute a certain amount to obtain a right to a certain share of the plant's services. Thus, if the plant operated 300 days a year, a country with yearly quantities of spent fuel taking 30 days to reprocess might be required to invest an amount equal to 10 percent of the initial capital costs to guarantee use of the plant for 30 days a year. The country would still pay utilization charges for the reprocessing, but its initial investment would serve to guarantee a stipulated access to the plant.

This allotment of investment shares is similar to what was used for the initial financing of Intelsat. Under the Interim Agreement signed in 1964 (Intelsat, 1964), the parties set up a consortium whose members contributed a share toward what was expected to be the cost of setting up a global satellite system –$200 million. The members owned the system in undivided shares, but the amount each country was to invest was determined by the extent to which it was expected to use the Intelsat space segment in 1968, the year the satellite system was expected to be operating.

The initial contribution need not, however, be related to expected or desired use of the facility. The financing of the enterprise provides a ready means of subsidization, if that should prove desirable, thus increasing the economic incentives for consumer participation. One or more supplier or nuclear weapons states might be prepared, as a matter of policy, to provide all or a disproportionate share of the initial capital. At a minimum, it would seem that the state of the supplier entity should be prepared to provide debt financing on terms at least as favorable as would be available to a national purchaser of the same facilities, and that would involve some concessionary element.

Decisions on all these matters would presumably be reached in the course of the negotiations establishing the enterprise and would be embodied in the constituent instruments.

The constituent instruments will probably establish only the guidelines for future decisions on follow-on financing, which can be made only when future needs and market conditions are better known. There are, however, several general provisions that should be made. The governing board, by majority vote, should have the authority to propose an increase or reduction in the amount of authorized capital. The recommendation would be adopted by the vote of the Meeting. However, if new shares are to be issued, two rules should be followed, except in the case of admission of new members, which should be provided for separately. First, each current shareholder should have the preemptive right to subscribe to new shares or quota subsequently issued in the proportion then held by that participant in shares or quota outstanding. This will prevent a dilution of an investor's interest by allowing each one to retain, after the new issue, the same proportionate interest in earnings, assets of the company, and voting power. Second, approval by a special majority of the Assembly should be required for any increase or reduction of capital which would result in changing the distribution of capital among the investors. The governments have a strong interest in the power distribution in the enterprise, so it is only natural that they should have a voice in any proposed major changes in that distribution. The Eurochemic Convention and Statute contained provisions for both preemptive rights and government approval.

Location

The site of the plant will have a decisive impact on all the major objectives —economic viability, nonproliferation, and security of supply. If, as may be the case, the enterprise is organized to take over an existing national facility, the siting decision will have been made for it. If not, the parties will almost certainly wish to agree in advance on the country, and perhaps even the general area within it, where the plant is to be located. These matters would be specified in the constituent instruments.

As suggested above, a supplier country would be optimal from the standpoint of nonproliferation, since it would already have reprocessing technology and know-how. On the other hand, it is possible that this alternative would be politically unacceptable to the consumers. Moreover, if, as is likely, the supplier would have the physical power (and conceivably, in certain circumstances, the motive) to interrupt operations or deliveries, the location might well be regarded by the consumers as inconsistent with security of supply.

It has sometimes been suggested that a remote island location would be desirable. But this might well involve serious security penalties due to extended transportation requirements and perhaps even the problems of defending against an organized raid. In any case, it may be difficult to identify a remote, sparsely populated island with the necessary resources and infrastructure—and even perhaps the space for spent-fuel and waste storage. Nor is such a location likely to be attractive as a long-term abode for the personnel of the enterprise and their families.

It may well be, therefore, that location in one of the consumer countries is indicated, especially for a regional facility intended to serve a relatively homogeneous geographic area. Ideally, the country should be one with a stable political and social structure, well-developed indigenous technical capability, and little independent incentive to acquire nuclear weapons. Important regional military powers or states with hegemonic ambitions should be avoided.

An important consideration will be the willingness and physical capacity of the country to receive and store spent fuel and radioactive waste. It is likely that special compensation or benefits would be provided for the performance of these functions.

Despite the concern that has been expressed, there is slight danger that the facility would be expropriated by the host country. Such a takeover, particularly by a state without nuclear weapons, would be a grave international incident, even more serious, for example, than withdrawal from the NPT or abrogation of a safeguards agreement. It would be likely, in fact, to trigger a major international crisis, and even reprisals or formal sanctions. The political inhibitions against such action are thus inherently severe. They could be strengthened even further by giving the facility extraterritorial status, which is by no means unusual for international organizations.

Various forms of harassment or interruptions of service by the host country are perhaps somewhat more likely than outright expropriation. The constituent instruments should provide for freedom of movement and appropriate immunities for the personnel of the enterprise and representatives of the participants. The status of the organization and its property with respect to taxation and judicial jurisdiction of the host country should be clearly specified.

Rights and Obligations of the Parties

For consumers the principal advantage of the multinational facility, and therefore the principal right, is access to reprocessed fuel. It is not a simple matter, however, to define the contours of that entitlement. The simplest formula would be that a country would have the right to the reprocessed fuel derived from its own spent fuel delivered to the plant. But, as will be seen, this arrangement—in which the participant gets back the same material it put in—may not fit well with the most efficient operating procedures.

An alternative would be to credit the participant with the energy equivalent of the recoverable material in its spent fuel, computed in accordance with an agreed formula. This approach would be particularly appropriate where the spent fuel was to be stored at the facility for a considerable period before reprocessing. Still another possibility, suggested above, would be to give the participant a claim on a specified portion of the annual capacity of the plant.

Under any arrangement, if the consumer had no current need for nuclear fuel, it would be permitted to sell or assign its entitlement.

As we have already noted, the consumer or its assignee should not be entitled to withdraw plutonium in pure or relatively accessible form in advance of need for it as power-reactor fuel. Otherwise, the multinational reprocessing plant would become, in effect, an international bomb-making establishment. Control over the release of the reprocessed fuel is thus essential for nonproliferation purposes, but it may be seen as a severe qualification on assurance of supply. In order to reconcile these objectives, the criteria for release should be specified with care in the constituent instruments together with the decision process for determining when the criteria had been met. These criteria would be based on current need for the reprocessed material as reactor fuel. The more objective they can be, the better the requirement of security of supply is met.

One way to insure that release decisions are made in a reliable and politically neutral manner is to vest this function in the IAEA. The reprocessed material, in the form of plutonium nitrate or oxide or fabricated fuel rods, would be placed in a depositary under the jurisdiction of the IAEA at or near the plant site until the consumer or its assignee requisitioned it. The IAEA would then apply the criteria specified in the constituent instruments to determine whether the requirement of need was met, perhaps with an appeal to the participating governments in the case of a negative decision. As already pointed out, the IAEA Statute gives it authority to perform these functions (IAEA: Art. XII, A5). It should be noted, also, that IAEA safeguards would attach from the moment the material was released from the depository.

In some cases the consumers might be prepared or required to forego any national reprocessing plants as an obligation of membership in the multinational enterprise. This might be the case, for example, in a regional organization, and it would appear clearly appropriate if the facility were established in connection with a nuclear-free zone. If the consumers were to accept this limitation, it would seem proper for the supplier members to assume a reciprocal obligation not to provide plants or relevant technology on a national basis to any of the consumer participants. Whether, in this context, it would be possible to obtain a broader agreement of the suppliers to sell only to approved multinational organizations is more doubtful.

Operational Questions

Design and Construction. At least four important decisions will occur in the design and construction phase: (1) selection of plant site; (2) selection of architect engineer; (3) selection of design; and (4) award of construction and supply contracts. Assuming that the host country has already been designated in the constituent agreements, these are all technical issues, though they are important enough not to be left to the discretion of management alone. The basic mode of decision should be by the governing board, on recommendation of the general manager. Some qualifications on this basic principle should be noted, however.

We have already suggested that the special concern of the host country with health and safety would probably carry with it a right to approve the plant design. Similarly, the host country would be a major participant in site selection.

Because of the proliferation aspects of these decisions, it may be desirable to limit participation of other consumer representatives. Of particular importance on this score is approval of the design of the plant, since this function could not be performed without access to sensitive technology and know-how. Similar considerations are applicable to the approval of specifications and contracts for the supply of sensitive items to be incorporated in the plant. Consumer representatives could readily participate in the establishment of general performance and financial constraints that the design would have to meet, in site selection and the choice of an architect-engineer.

Supply and construction contracts should be awarded on the basis of competitive bidding. Some countries participating in international enterprises, particularly the Europeans, have put forward a principle of *juste retour,* under which procurement is allocated in proportion to investment. This should be resisted. It is significant that Intelsat beat back a determined effort to include such a principle in its Definitive Arrangements (Intelsat, 1974). And even Eurochemic, which had a dozen design offices involved in design studies to maximize the spread of design know-how, awarded supply and work contracts after international invitations to tender. Subject to the qualification about sensitive items mentioned above, there is no reason why contractors from consumer countries and even from nonparticipants should be excluded from the competition.

Staging of Co-Located Facilities. For convenience, this paper has primarily addressed the fuel reprocessing function. As noted above, however, other services could be provided by a multinational fuel center, including spent-fuel storage, fuel fabrication, and long-term waste disposal. Indeed, a spent-fuel storage facility might be the first to be built in a multinational center. The spent fuel could be stored there until reprocessing became economically attractive, thus relieving the participants of the need to build additional national pools. Moreover, it would be a useful confidence-building activity. For the reasons already given, fuel fabrication must be provided if the plutonium recovered in reprocessing is to be returned to the consumers in any form. Long-term waste disposal could provide a real incentive for participation since it is a difficult, hazardous, and potentially unpopular service that no country has really faced up to. Countries without suitable geological formations for long-term storage would be especially attracted by a reprocessing enterprise that also performed this function.

Given this range of services, when and in what order should the facilities be built? If spent-fuel storage only is to be provided first, the parties must agree if and when to build the reprocessing plant. They might provide in the constituent instrument that, in principle, the enterprise would store until reprocessing became economically attractive, with a special planning group made up of government and industry representatives to review the situation periodically. This group would recommend when to begin reprocessing. Both the

Assembly and the Meeting would concur in the approval of the recommendation. Dissidents should have the right to opt out of the enterprise at that point if they opposed the reprocessing decision.

Once the decision to reprocess is made, whether in conjunction with a spent-fuel storage facility or by itself, it will be necessary to decide about fuel fabrication and long-term storage. If the consumers want their plutonium back, then fuel fabrication is necessary since, as noted, it would be absurd to send pure plutonium back to the consumer. However, if they were willing to store the plutonium at the plant or to sell or exchange it with approved recipients, then fuel fabrication could be excluded initially but perhaps provided later by concurrent vote of the Meeting and the Assembly. The constituent instruments should so provide.

Provision of long-term waste disposal will depend in large part on the attitude of the country likely to host the storage facility. Since long-term wastes have no economic value, it appears likely that participants would utilize such storage facilities if they were available. If a host for such a storage facility —which need not necessarily be the same country in which the plant is located— could be found, then a special agreement outlining the arrangements should be worked out between that country and the other participants. If, on the other hand, a satisfactory storage host cannot be found, the constituent instruments should make it clear that the wastes are to be returned to the country of origin. The details of transportation and timing could be worked out by the governing board.

Personnel. One of the first questions will be the selection of personnel for the facility. This should not be a political matter. Operatives should be selected on the basis of technical competence and responsibility. The general manager, as the management head with the greatest technical and industrial competence and who will have control over operations, should have responsibility for selection of personnel. Due to technology-transfer considerations, the general manager might be instructed to consider applicants from only certain countries for certain sensitive positions.

National quotas should be avoided for at least two reasons. First, they emphasize the national rather than international affiliation of the person. This is directly contrary to the plant loyalty desirable in the staff. Second, it is costly. A quota system may not only result in the selection of less qualified persons; it can also result in overstaffing of the facility. Eurochemic, with a capacity of 108 MTU/yr had 350 employees, while the West Valley plant of Nuclear Fuel Services, with a capacity of 300 MTU/yr, had 170 employees (USAEC, 1968: 167, 179). One wonders how many of those 350 employees were hired to satisfy a directive to recruit "from all the participating countries on as balanced a basis as possible" (OECD, 1962/64: 2). Thus, while national origin could provide a basis for choice between two equally qualified and necessary persons, balanced national representation should certainly *not* be an excuse to create make-work positions.

Batching, Scheduling, and Queuing. Because of the danger that critical masses of plutonium might accumulate during the reprocessing cycle, spent fuel is reprocessed in batches, rather than in a continuous flow. This gives rise to the question of what size batches should be processed at one time. In practice the choice is whether to reprocess (1) only one country's fuel at a time, even if the economically optimum batch size is greater than the national batch size; (2) to sequence two batches of similar fuels in a continuous campaign; or (3) to process different fuel batches as closely in succession as practical. Economically, the use of batches geared to plant capacity is preferable because it leads to fuller utilization of the facility. The use of national batches could result in underutilization, seriously affecting the realization of economies of scale. For example, the United States Atomic Energy Commission (USAEC) calculated in 1970 that a 300-MTU/yr plant operating at 100 percent capacity could reprocess fuel at $60/kgU. A 1500 MTU/year plant operating at 20 percent capacity could reprocess at $98/kgU; the same plant at 50 percent and 100 percent capacity could reprocess at $45/kgU and $27/kgU, respectively (USAEC, 1970: 263). These figures are no longer valid in absolute terms, but they still indicate the impact of underutilization on reprocessing costs in large plants.

The main disadvantages of processing different fuel batches in close succession are the mixing of fuels of different isotopic composition and the loss of precision in closing material balances on individual batches. However, both the USAEC and the preliminary studies at Eurochemic indicate that the cost advantages outweighed these disadvantages, because of the costs of the shutdowns in a larger plant that did not reprocess similar fuels in close succession. The Barnwell plant, for example, would require weekly shutdowns if it processed by consumer batches (Gallagher, 1975: 112–15).

If a consumer's entitlement was to receive the reprocessed material derived from its own spent fuel, the problems of material balances and mixing of fuels might assume considerable importance. Under such a regime, national batching might be indicated, despite the economic penalties. There is a potential vulnerability in this sytem, however, for if the nationality of each batch is known, it might open the way for politically motivated shut-downs. This would be more difficult if all similar fuels were kept in a single common pool after the initial physical and analytic measurements were taken. These considerations, together with the efficiency advantages, militate strongly in favor of batching on a basis related to the capacity of the plant rather than the origin of the spent fuel. It follows, also, that the consumer would not be entitled to the reprocessed material derived from its own fuel, but to an equivalent energy value.

Under the economic batching method, it would be necessary to accumulate spent fuel in the amount of the optimal batch each time the plant is ready for a new charge. This would not be difficult if there were a backlog of spent fuel that could be drawn on smoothly and continuously to supply the reprocessing plant. The backlog would be replenished by new deliveries of spent fuel, more or less at the convenience of the consumers. The contribution of such a pool to the efficient operation of the plant is another good reason for starting out with spent-fuel storage, before the reprocessing plant comes on line.

If there is no such backlog, the manager will have to arrange with consumers to schedule deliveries of spent fuel so as to accumulate an optimal batch at the appropriate times. Efficiency of operations will depend on meeting this schedule, so a consumer that failed to meet its delivery commitment should probably be subject to some kind of penalty, under rules and regulations prescribed by the governing board.

The effectiveness of the scheduling will be heavily sensitive to the accuracy of the management's estimate of the amount of time during the period covered by the schedule that the plant will be out of service for cleaning and repairs. Overestimation has the effect of understating the capacity of the plant. Lags will develop. Optimal batches will not be on hand in time. And some portion of the capacity will not be used. Underestimation of down time, on the other hand, will generate queues, and consumers will experience unnecessary delays in reprocessing.

Scheduling of spent-fuel deliveries would seem to be an ordinary efficiency-maximizing function of management. The picture changes slightly if we consider the possible desire of the consumer for delivery of reprocessed fuel by a certain date (assuming that the release criteria have been met). This matter also might be remitted to ad hoc arrangements between the consumer and management. However, a general right to delivery within a fixed period may be seen as an aspect of security of supply, in which case consumers might want greater assurance as to maximum delivery time than the opportunity for case-by-case negotiation. But the achievable maximum delivery time would be heavily affected by the actual performance of the plant, which would itself be expected to vary over time. As a result, it would probably be impracticable to specify the maximum in the constituent instruments, although they might contain guidelines and, perhaps, an extreme outer limit. The actual determination of maximum delivery time would be made by the governing board, and altered from time to time as necessary in the circumstances.

The imposition of such a constraint may affect the efficiency of operations in two ways. If lags develop in the schedule, the plant may have to reprocess fuel, even though an optimal batch has not yet accumulated, in order to meet the consumer's required delivery date. If there is a queue, it may be impossible to meet the delivery date for some consumers. In that case the enterprise would presumably become liable for penalties, under rules and regulations prescribed by the governing board, in the same way as the consumer who fails to make a scheduled delivery of spent fuel.

The prospect of queuing, whether because of scheduling mistakes or otherwise, raises the question of the order in which the consumers in the queue will be handled. A first-come-first-served rule would seem right, to a first approximation, but exceptions may be necessary where a late arrival has urgent need for reprocessed fuel. The elaboration of the rule and the criteria for exceptions should be a function of the governing board, with implementation left to management. Provision might be made for an appeal to the board by a consumer adversely affected by a management decision.

As already noted, without national batching the consumer's entitlement will be expressed in terms of an energy equivalent, translating tonnes of spent fuel delivered into tonnes of mixed-oxide or UO_2 fuel. The Nuclear Fuel Services plant in West Valley, New York, reprocessed but did not refabricate fuel from several reactors, including the Yankee-Rowe and Consolidated Edison–Indian Point. The plant used the following accountability techniques:

> To account for the quantities of special nuclear material contained in the fuels, NFS made physical and analytic measurements of the dissolved fuel solutions, recovered product solutions, and waste streams as required by its contract with AEC. The AEC verified these measurements by an on-site program of surveillance and independent analytic measurements to ensure equitable financial settlements between the Commission and NFS for the reprocessing service and between AEC and Commonwealth Edison and Yankee for the value of the uranium and plutonium contained in the fuels [Abrams, 1969: 187].

In addition, the recipient weighed each bottle of plutonium product solution as received, to ensure that the previous measurements were reasonable (Abrams, 1969: 192).

Although the exact nature of suitable accountability techniques for a multinational plant is yet unclear, it would seem that the IAEA could play a large role as an independent auditor. IAEA inspectors will already be checking measurements of inputs and outputs of special fissionable materials for safeguards purposes, and there seems little reason why they could not also be the official accountability experts.

If batches are mixed and the fuel refabricated, it will be necessary to derive an energy-equivalent coefficient for each fuel type or reactor. The constituent instruments could establish the guidelines for determining these coefficients, but a particular coefficient could not be derived until the fuel type in question was run through the plant and appropriately measured. This task can only be done by the technical operators, under the supervision of the general manager and IAEA inspectors. If a country wished to challenge its assigned energy-equivalent coefficient, it could have the right to new measurements by a different IAEA team, perhaps on condition that it pay the costs of the second reading.

For those many other decisions affecting technical operations not considered above, it appears best to leave them to the general manager, who would, of course, solicit views from the staff. But the general manager's decision should be final. If certain actions are unreasonable, the governing board would have the power to remove the general manager by a three-fourths' vote. Barring gross improprieties, however, the general manager should be given broad discretion to manage the plant in a nonpolitical, technical, and industrial fashion.

Pricing. Price-setting dramatically illustrates the complex interrelation of decision levels in a multinational reprocessing enterprise. Basic pricing norms will be set in the constituent instruments, not only because of the relationship to security of supply, but because they crucially affect the overall benefits to be derived by the participants from the enterprise. Subject to these basic rules, broad price policies should be established from time to time by the Meeting or the Assembly. Actual price-setting should be the function of the governing board on recommendation of the general manager.

If the use of the facility is proportionate to investment share, much of the pricing problem could be short-circuited by charging each user a pro-rata share of operating costs. This is basically the Intelsat method. It seems unlikely that investment and use would be closely correlated in a reprocessing facility, however, unless all the participants were from developed countries. Even then the frequent readjustment of investment share to correspond with actual use, as provided in the Intelsat agreements, would probably be unacceptable in the reprocessing context.

If this alternative is foreclosed, price-setting policy would have to take account of two potential groups of customers—participants and outsiders—and two possible price indicators— average cost and market price—assuming that there is an effective market for the services provided by the plant or their equivalent (enrichment, for example). The constituent instruments should establish the conditions under which services will be available to outsiders, presumably only after all claims by participants for the relevant scheduling period have been provided for. The decision that these conditions have been satisfied could be made by the governing board.

The constituent instruments should also enumerate the components of average cost, including—in addition to current operating expenses—provision for interest and amortization of debt, reserves, and a reasonable return on investment (unless the investors are prepared to subsidize the operation by accepting less than a reasonable return). The actual figures for the reserves and return on capital components should be set by the Meeting on recommendation of the governing board.

The constituent instruments might also include guidelines for determining whether an effective market alternative was available for reprocessing services or their equivalent. But again, the derivation of the market prices prevailing at any given time, which would be a complex process, might be determined by the Meeting, or it might fix more precise rules, with their application left to the governing board.

Pricing policy must provide both for the case when average costs are lower than the market and the case when they are higher. When average costs are lower, it would appear appropriate to give the participants the benefit of a price based on average costs, although the case might be different if the investors were primarily interested in the profit-making potential of the enterprise. In any event, there seems no good reason to charge outsiders less than market price, unless the gains in terms of capacity operation warrant some concession.

When average costs are above market, outsiders would presumably be unwilling to pay the average costs. Nevertheless, if capacity were available, it would pay the enterprise to provide services to outsiders as long as market price exceeded variable costs. The situation with respect to the participants would be no different, unless they were bound to use the services of the plant under some kind of take-or-pay arrangement or in order to maintain their entitlements. If there is any such obligation, its nature and extent should be spelled out in the constituent instruments.

It will be seen that there is enough potential for variation in the configuration of these elements so that responsibility for developing and executing policy in response to them should be vested, respectively, in the Meeting and the governing board. These functions should be exercised subject to an overriding provision in the constituent instruments requiring equal treatment as between participants. A participant that regards itself as discriminated against should have the right to bring its complaint to the Assembly, which would have authority, perhaps by a special majority, to order the practice terminated.

Rules for the distribution or retention of earnings should also be contained in the constituent instruments. The actual decision to distribute or retain earnings seems best left to the board of directors. Investors will have vested interests in these rules, so it would be best if they could be changed only by a special majority of the Meeting. If a more flexible amendment procedure appears desirable, then capital contributors might be given an option to sell out if outvoted on a crucial issues, such as the rules for distribution of earnings.

Safeguards. Hopefully, many safeguard measures will be built into the design of the plant. In fact, one of the important advantages of a multinational approach is the enhanced opportunity it may provide for involvement of the IAEA at the design stage. The initial standards for safeguards will be largely the responsibility of the IAEA and the supplier of the facility. The formal IAEA safeguards agreement must be concluded between the supplier, the

IAEA, and the recipient before the reprocessing equipment is sent to the plant site. Since access to the plant will be across the territory of the host country, it, too, must be a party to the safeguards agreement.

In addition, the government-level agreement should include an undertaking by all participating states to do everything possible to ensure adequate safeguarding. All important questions relating to safeguards should be referrable to the Assembly by any staff member, manager, director, or investor. A representative of the supplier government should be present at any Assembly called to consider a safeguards question. Under no circumstances should any organ of the enterprise be given the authority to expel any IAEA inspectors or to repress any of their findings. Any technical operator should be given assurance that reports made will be distributed anonymously, and that individuals will be protected from any management or board actions designed to intimidate or punish for any reports of possible violations to the Assembly. The safeguards area is one in which, because of the sensitivity to nonproliferation concerns, consideration should be given to requiring the separate concurring vote of supplier members for any action by the relevant decision organ.

CONCLUSION

The foregoing is hardly more than a catalog of some of the major issues that would have to be faced in negotiating the institutional structure of a multinational nuclear-fuel center. It is a formidable list. Nevertheless, we can conclude from this review that none of the organization problems is inherently insoluble. Difficulties are exacerbated by the large number of questions, their interrelations, and the complexity of the assumed structure.

It should be recalled, however, that our assumptions posed the most difficult case. A reduction in membership and in the ambitious scope of the enterprise, at least at the outset, would greatly reduce the complexity of the organizational structure and the difficulty of negotiations. The analysis suggests that multinational fuel-cycle activities should start out more modestly, perhaps only with joint appraisal by a relatively few countries with existing geographic or economic connections. If operations are contemplated it would seem that the first step should be joint arrangements for spent-fuel storage, with the decision to go forward to more elaborate activities deferred. This approach would not only be simpler and permit the parties to gain experience working together, but it would have the virtue of delaying reprocessing until it was clear that there was a real need for it. The IAEA seems to be moving toward this approach to the matter.

Even on this reduced basis, the negotiating task would not be easy. The key, of course, to overcoming difficult technical problems of institutional structure is political will—the genuine commitment of the participants to the aims and values of the enterprise. This suggests that any effort to cajole—not to say coerce

–participation in a multinational fuel-cycle enterprise would be wholly misplaced. A reluctant partner would have available an infinitude of points and issues to create plausible, irritating, and ultimately defeating delay and complication in the negotiating process. Only assent freely given in the perception that the enterprise really serves the interest of the countries involved will be able to surmount the many institutional problems that will inevitably arise.

REFERENCES

Abrams, L.A. "Fuel Reprocessing–Commercial Experience," *Reactor and Fuel-Processing Technology* 12 (Spring 1969).

Fligler, Carlos. *Multinational Public Enterprises.* International Bank for Reconstruction and Development, 1967.

Friedmann, Wolfgang G. "International Public Corporations," *Modern Law Review* 6 (December 1943).

Gallagher, James M. *Nuclear Fuel Reprocessing Expansion Strategies.* Technical Report SOL 75-11. Stanford, Ca.: Stanford University Department of Operations Research, May 1975.

IAEA. *Statute of the International Atomic Energy Agency.* 8 U.S.T. 1093; T.I.A.S. No. 3873; 276 U.N.T.S. 3 (1957).

Intelsat. *Agreement Establishing Interim Arrangements for a Global Commercial Communications Satellite System.* 15 U.S.T. 1705; T.I.A.S. No. 5646 (1964).

Intelsat. *Agreement Relating to the International Telecommunications Satellite Organization (INTELSAT).* 23 U.S.T. 3813; T.I.A.S. No. 7532 (1973).

Organization for Economic Cooperation and Development (OECD) and European Nuclear Energy Agency. *Second Report on the Activities of the Eurochemic Company, 1962–1964.*

Sundström, G.O. Zacharias. *Public International Utility Corporations.* Leiden: A.W. Sijthoff, 1972.

Tannenbaum, Jeffrey A. "White Elephant? Big Plant to Recycle Nuclear Fuel Is Hit by Delays, Cost Rises," *Wall Street Journal,* February 17, 1975, p. 1.

United States Atomic Energy Commission (USAEC). *Nuclear Industry 1968.* Washington, D.C.: U.S. Government Printing Office, 1968.

——. *Nuclear Industry 1970.* Washington, D.C.: U.S. Government Printing Office, 1970.

Chapter 12

Some Experiences in Formation and Operation of Multinational Uranium-Enrichment and Fuel-Reprocessing Organizations

C. Allday

Urenco/Centec is a tri-national organization involving British, Dutch, and German companies, and is devoted to development and exploitation of gas centrifuge technology for enrichment of uranium. The organization involves two companies, Urenco Ltd. and Centec GmbH, but is run as a single entity.

United Reprocessors GmbH is a similar international venture for fuel reprocessing. The partners in this case are British, French, and German companies.

Both organizations were formed because it was recognized by the respective shareholders that collaboration was essential in order to establish viable industries in these two important sectors of the nuclear fuel cycle. The motives were essentially commercial, but political considerations, particularly those concerning nonproliferation and therefore of safeguarding dissemination of technological information, have had a major influence on the policies of both organizations. Urenco/Centec was in fact set up under the aegis of an international treaty—the Treaty of Almelo*—while the activities of United Reprocessors are closely followed and in political matters influenced by the three governments concerned.

This paper describes the history of formation of each organization and the progress during the first few years of their existence. It is hoped that the information will be of interest and some value to those considering other multinational organizations, and particularly the so-called regional reprocessing centers which are the subject of the conference for which the paper has been written.

*The Treaty of Almelo was an agreement between the United Kingdom of Great Britain and Northern Ireland, the Federal Republic of Germany and the Kingdom of the Netherlands on collaboration in the Development and Exploitation of the Gas Centrifuge Process for producing Enriched Uranium—Cmnd 4315—Her Majesty's Stationery Office, London.

URENCO/CENTEC CENTRIFUGE ENRICHMENT ORGANIZATION

Background

Investigation of the centrifuge method of enrichment of uranium in its 235 isotope had been carried on in several countries, particularly the United States and Great Britain, from the earliest days of atomic energy. However, the technique was not recognized as having practical value and commercial potential until after publication of the work of Zippe, et al. (1958) in Germany. Thereafter work in several countries was intensified. For instance, in Great Britain a significant development program was started in 1961, and by 1968 it had been decided that the centrifuge process was both technically sound and practicable, and had the prospect of being significantly more attractive economically than the established diffusion process.

The work had been carried out in utmost secrecy in all countries, but through diplomatic channels it became evident that development was proceeding in parallel in Germany, Holland, and Great Britain. The industrial interests concerned recognized that the development would be both lengthy and costly and that the market could be very competitive. Governments recognized the desirability of keeping the development under firm political control, and so it was agreed on both the industrial and political planes that collaboration was desirable. With the assistance of the industrial interests, governments negotiated the Treaty of Almelo which was signed in 1970 and ratified by all three countries in 1971. It was under the auspices of this treaty that Urenco Ltd. and Centec GmbH were formed in 1971.

The Treaty of Almelo

The basic provisions of the Almelo Treaty are:

1. The three countries would collaborate in the centrifuge enrichment of uranium and in the development and manufacture of centrifuges.
2. Joint industrial enterprises would be set up with cross-participation from each country.
3. A political Joint Committee would be set up to exercise appropriate government control of sensitive political issues.
4. All information concerning the process would be kept secure in the interest of the nonproliferation of nuclear weapons.

Accordingly, Centec GmbH was formed as a German-based company with the responsibility of developing centrifuge technology and establishing centrifuge machine manufacture and enrichment plant construction. Shareholders are: British Nuclear Fuels Limited (BNFL) in Great Britain, Ultra-Centrifuge Nederland (UCN) in Holland, and Gesellschaft fur Nukleare Verfahrenstechnik (GnV) in Germany. Similarly, Urenco Limited was formed in Great Britain with the same British and Dutch shareholders as in Centec, but with URANIT the German shareholder. Details of the composition of the various shareholders is given in Figure 12-1.

The reason for having two companies and for the difference in shareholder in the German case was the German conviction that for prudent commercial reasons development and manufacture of the process and plant should be kept separate from the operation of plant and sale of enrichment. In the Dutch case, UCN was an amalgam of both sets of industrial interest, and in the British case BNFL was already concerned with both aspects.

It was the original intention that both enterprises should have a high degree of autonomy and independence. Thus Centec would control R&D programs and set up independent machine-manufacturing facilities and control its own plant design organization. Urenco would buy plants from Centec and would operate them and sell the product. As described below, this was fairly quickly recognized as unrealistic and impracticable under the prevailing circumstances and on a short timescale.

Organization Developments

The proposed structures of Urenco and Centec had been formulated in advance of comparison and cross-knowledge of the respective technologies being developed in the three countries. It had been assumed that it would quickly and fairly easily be possible to identify a single line of development for exploitation on a centralized basis. This did not prove to be the case, and it was soon recognized that while considerable concentration and narrowing down of development objectives was possible and that great benefit was derived from the interflow of information, there was more than one well-established technology that merited continuing development. This, and the vast amount of money involved in investment decisions, led to the conclusion that a looser structure was required, at least for the time being. The change, which was introduced in 1974, works well in that it facilitates excellent technical and commercial collaboration. It is outlined and illustrated in Appendix 12-A. Under this arrangement each shareholder is responsible for its own development program but judges it in the light of complete and free exchange of information, and as controlling shareholder in each national enterprise determines the extent of major investment in plant it is prepared to make and what technology it will use.

Under the arrangements outlined above, the shareholders exchange all their technical information and forward development plans, economic studies, actual costs of plant, and the like. Marketing is done jointly for them by Urenco with mutually agreed terms and conditions. Fourth-country exploitation of the technology is carried out by the central organization, and all the shareholders have given up their rights to act independently. As noted below, sale or passage of technological information is politically controlled by the Joint Committee. The

Uranit Uran-Isotopentrennungs-gesselschaft mbH	BNFL British Nuclear Fuels Ltd.	UCN Ultra-Centrifuge Nederland NV	GnV Gesellschaft fur Nukleare Verfahrenstechnik mbH
20 % Farbwerke Hoechst AG 40% Gelsenberg AG 40% NUKEM GmbH	100% United Kingdom Atomic Energy Authority	55% Staat der Nederlanden Reactor Centrum Nederland (the Dutch state-owned Atomic Energy Authority) 10% Philips Gloeilampen -fabrieken NV. 7.5% Rijn-Schelde NV 10% Shell Kernenergie NV 10% Staatsmijnen 7.5% Verenigde Machine-fabrieken NV.	50% Interatom GmbH 50% MAN AG.

Figure 12–1. Shareholders of Urenco and Centec.

central organization is based at Marlow, England, and employs 46 staff, of whom a quarter are German or Dutch. There can now be no doubt that Urenco/Centec has the potential to become a major supplier of both enrichment plant technology and of enrichment.

Political Control

The political aspects of the activities of Urenco/Centec and, through the Shareholders Agreement, the activities of the industrial shareholders in the centrifuge field are controlled by the Joint Committee. This has government representatives from all three countries concerned, and the chairmanship rotates on an annual basis. The Joint Committee, is required by the Treaty of Alemelo, and has as its main aim the supervision and control of the dissemination of centrifuge technology information as well as coordination of the application of funds provided by the three governments. (At present the R&D programs in all three countries benefit from some degree of government support.)

To these ends the Joint Committee controls classification policy with respect to centrifuge information; accepts responsibility to ensure that the policies of the companies safeguard enriched uranium against diversion; controls dissemination of information to fourth parties, particularly by requiring Urenco/Centec to submit for its approval all proposals for dealings with fourth countries; and approves R&D programs and investment proposals of the enterprises and shareholders. These arrangements have proved to work extremely well, and with a large proportion of finance in each of the three countries being provided by governments, close and generally harmonious relations between the shareholders and governments have developed.

General

Urenco/Centec is a successful multinational organization in the centrifuge field. An effective system of working has been evolved which satisfies the normal industrial and commercial aspirations of the shareholder and yet provides appropriate political control to ensure the proper safeguarding of the technology and products. Such a system inevitably involves sacrifice of certain freedom of action by the industrial shareholders on the one hand and on the other a recognition by governments of the need for speedy mechanisms for consultation and decision-taking. Such amalgamation of industrial and political interests on a multilateral basis is not easy, but the considerable success achieved so far in the case of Urenco/Centec demonstrates that industrial cooperation with political control is possible. The final success of Urenco/Centec depends on many factors —technology (which is established), economy of process (which appears to be more than competitive), the market (which while hesitant and uncertain in the near term seems assured in the longer term), and continuing efficient international collaboration at both industrial and political levels, which as reported in this paper is standing up well to the practical test of time.

UNITED REPROCESSORS GmbH

Background

Reprocessing is the final step in the nuclear fuel cycle. Irradiated fuel after releasing energy in the reactor is taken to a reprocessing plant where it is first physically treated to expose the uranium metal, or oxide as the case may be, from the metallic cladding of the fuel elements. The released fuel is then chemically treated to separate the uranium, plutonium, and radioactive fission products, the last two having been formed during the period in the reactor.

Reprocessing was developed by the nuclear powers as part of their defense programs, and large reprocessing plants were, therefore, established in the United States, Great Britain, and France (as well, of course, as in the Eastern world). With the development of nuclear electrical power generation, commercial reprocessing plants were established in the same three countries, and an internationally owned plant–Eurochemic–was built in Belgium. Several other countries have small pilot or experimental plants. During the 1960s there was a surplus of capacity over demand (mainly because of the reduced military programs), and the potential capacity situation with new plants under construction or planned in at least the United States, Great Britain, France, and Germany, appeared to indicate an even larger surplus. For Europe a special study group set up by Foratom recommended that this situation be recognized and that there should be international collaboration to rationalize investment in order to establish a viable industry. Consequently, the British, French, and German reprocessing interests (BNFL, CEA [now COGEMA], and KEWA) joined together and formed United Reprocessors GmbH.

Objectives and Functions

United Reprocessors provided joint marketing of existing capacity and an organization under which the shareholders could rationalize their future investments according to agreed rules. The company also became the repository for all the technical know-how of the shareholders, who had agreed to make it freely available to each other and sell it to fourth countries only through United Reprocessors. The arrangements do, however, allow for the shareholders independently to supply small, laboratory-scale reprocessing plants. Before transactions above that limit occur, unanimous agreement of the shareholders is required, and in practice the approval of the individual governments has to be obtained by the respective shareholders. Relations with government have worked well.

The Changed Reprocessing Scene

During the last few years, the overall position of reprocessing has changed significantly. Existing reprocessing plants have encountered technical difficulties, and construction of new plants has been considerably delayed. Thus at

Morris, Illinois, GE has experienced serious difficulties in the commissioning of their plant, which was designed using advanced process techniques, and the plant is now likely to be abandoned. Similarly, the NFS plant at West Valley, New York, is closed, and the AGNES plant at Barnwell, in South Carolina, has been seriously delayed, mainly because of licensing difficulties. In Great Britain, an accidental release of a relatively small amount of airborne Ruthenium 106 in the head-end section of the plant, which breaks down oxide fuel, occurred in 1973, and the plant has been closed ever since. The necessary modifications to prevent recurrence are relatively straightforward but are time-consuming and expensive on a plant which has already operated with highly radioactive material. This plant, which was originally designed for 400 tonnes/year throughput of oxide fuel, was to be expanded to 800 tonnes/yr, but this will not now be done, and a completely new plant of 1000 tonnes/year, designed to modern standards of safety and radioactivity release specified by the appropriate authorities, is to be built. In France a 400 tonnes/year head-end plant has been constructed on the La Hague site and is presently being commissioned. It is intended to increase its capacity progressively to 800 tonnes/year by 1980. In Germany, KEWA has not yet obtained a license for the construction of their planned 1500-tonnes/year plant, and the original date of operation of 1981/2 will not be met.

The costs of new plants have escalated dramatically under the influence of inflation and of tightened environmental, security, and health-physics requirements.

The United Reprocessors' Agreement and collaboration has proved to be remarkably flexible in these changing circumstances. The fact that each partner is responsible for the construction of plants in its own country has enabled plans to be changed without undue delay. The sharing of technology has worked to the advantage of all the shareholders, and the coordination of the R&D programs will ensure that the designers of the new plants will be provided with the best information at minimum cost and duplication of effort.

Under the new circumstances there is clearly going to be a shortage of capacity in the 1980s, and substantial quantities of fuel will have to be stored in ponds prior to eventual reprocessing. United Reprocessors' shareholders are planning their capacities of both storage ponds and reprocessing plants accordingly, but will not invest in plants for external business without long-term commitment by the customers, together with substantial prepayments to help finance construction of the plants.

Environmental Effects of Reprocessing

The foreseen expansion of nuclear reprocessing facilities in Europe has given rise to some concern about the possible effect on the environment. Environmental and resource limitations do exist on the scale of activities on any individual site, but there is no doubt that the planned expansions are achievable without a significant increase in the discharges to the environment. This is because of the developments in reprocessing technology which will be available for future plants. The main items are as follows:

1. *Solidification of high-level liquid waste.* Existing reprocessing systems are designed to ensure that more than 99 percent of the fission product activity is contained in a very small volume of liquid waste (0.25 m^3/t fuel). This is stored in high-integrity tanks which require surveillance, and spare tanks have to be available at all times. Processes to convert this waste into an inert, solid form are now being developed to allow this waste to be safely stored in a system which will require minimum surveillance and maintenance over long periods of time. The solidified waste will also be suitable for transport when such procedures have been introduced.
2. *Salt-free processes.* With present reprocessing techniques, about 60 m^3 of medium-active liquid waste containing about 10 Ci/m^3 are produced per tonne of fuel processed. In the future, new processes will result in lower salt contents of this category of waste enabling a high level of concentration to be achieved by evaporation. Resultant high-activity liquors will then be suitable for storage with the high-level wastes described above.
3. *Gaseous waste treatments.* A process for the removal of I-129 from gaseous waste is now under development and will be included in the new reprocessing plant to be operational in the eighties.
4. *Plutonium-contaminated materials treatment.* The following techniques are being developed to ensure effective management of this type of waste:
   a. Incineration.
   b. Recovery of plutonium from incinerator ash by acid leaching.
   c. The incorporation of leached ash into a ceramic matrix.
   d. Storage of ceramic blocks, after packaging, in man-made structures.
5. *Ultimate disposal of wastes.* The safe disposal of immobilized solid radioactive wastes will eventually become important, and this problem is being tackled on a worldwide basis. Practicable alternatives include disposal into stable geological formations or into the deep ocean. Possibility of retrieval for inspection would be required during the development of either of these options. A German concept involves disposal in deep saltbeds directly beneath a fuel-cycle park. In France, the main interest is in storage in granite or saltbeds. The U.S. program includes a national survey of multiple geographic locations and the construction of a demonstration underground storage facility in dry bedded salt to store plutonium-contaminated wastes.

General

United Reprocessors is an effective organization providing for collaboration between the main European reprocessors. It is not a plant-owning or investment company; it employs only six people under a *Geschaftsfuhrer* (general manager).

However, it has demonstrated the ability of commercial organizations, some government-owned and some privately owned, to collaborate and provide the basis for an orderly and rational development of the industry. Political control of activities, particularly those affecting return of fission products to customer countries, and dissemination of reprocessing know-how, is effected by the individual governments acting directly on the respective shareholder. Again, coordination of activities and implementation of uniform policies has so far been effective.

CONCLUSIONS

Three main points emerge from the experience gained in the formation and operation of these two organizations, which may be particularly relevant to the consideration of multinational regional reprocessing centers. It has become clear that the time scale involved in the formation and operation of these organizations is protracted partly because of the fact that more than one shareholder and government are involved, but also because some of the shareholders themselves have shareholders who have to be consulted on matters of particular importance.

It has been found that very careful consideration has to be given to the extent of the autonomy and independence of the joint organization. For Urenco/Centec and United Reprocessors it has proved to be more practicable for the shareholders to take the role of working partners.

Finally, when several technologies of virtually equal merit exist within a joint organization, it is very difficult, if not impossible, to make the decision to develop only one of them. In the case of Urenco/Centec each shareholder is responsible for its own development program but judges it in the light of complete and free exchange of information.

APPENDIX 12-A. THE STRUCTURE OF URENCO/CENTEC AND ASSOCIATED ENTERPRISES

The present structure of the tripartite centrifuge enrichment collaboration comprises the two companies (Urenco Limited and Centec GmbH) and two partnerships or enterprises (Urenco [UK] and Urenco [Nederland]). The involvement of the British, Dutch and German shareholders in these is illustrated in Figure 12-2.

Urenco Ltd and Centec GmbH

These two companies provide the central services required by the collaboration. They are controlled by their board and *bierat* respectively, but there are regular joint meetings to discuss mutual problems. The activities of both companies are now based at Marlow, England, under the direction of a common management. The principal functions of the two companies are:

Urenco Ltd

—To provide a central marketing service as the agent of the production enterprises.

—To assist in mutual scheduling of production.

—To represent the shareholders in respect of investment in enrichment projects with parties outside the tripartite enrichment organization.

Centec GmbH

—To coordinate the research and development program, and to facilitate the cross-flow of technical information between the shareholders.

—To represent the shareholders in respect of dealings with parties outside the tripartite enrichment organization for the sale or licensing of technology or of enrichment plant or equipment.

The Production Enterprises

The principal business of the enterprises is the design, construction, ownership, and operation of centrifuge plants for enrichment of uranium.

There are two enterprises, formed as partnerships with participation shares as shown in Figure 12-2.

URENCO (UK) owns factories at Capenhurst, England and is managed by BNFL as its majority partner.

URENCO (NEDERLAND) owns factories at Almelo, Holland, and is managed jointly by its majority partners, URANIT and UCN.

The managing partner (or partners) of each enterprise is (are) responsible for the investment decisions of the enterprise and for design and operation of the plants. The minority partners are, however, kept fully informed of the affairs of the

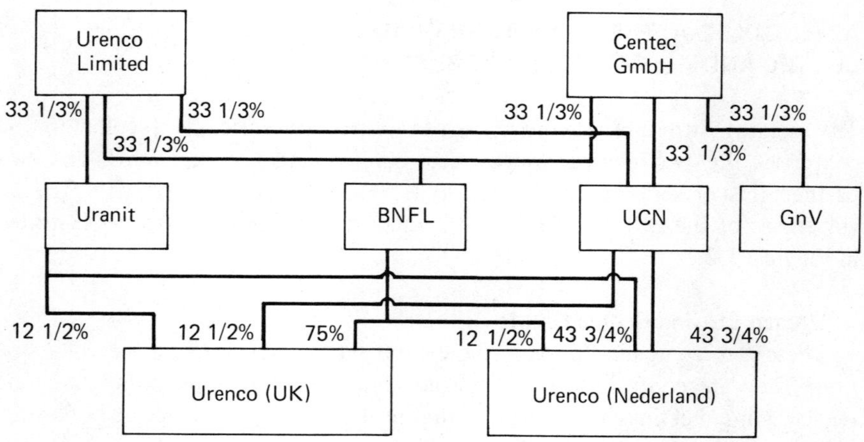

Figure 12-2. Urenco/Centec: Basic Organization.

enterprise and are able to express their views at regular meetings of the enterprise board. Business is sought by Urenco Limited against terms and conditions of sale determined by the unanimous agreement of its shareholders and is concluded on behalf of the two enterprises. The enterprises are entitled to shares of the business on the basis of one-third to Urenco (UK) and two-thirds to Urenco (Nederlands) (in recognition of the involvement of both Dutch and German shareholders). If one enterprise does not wish to undertake its full share of the business, it may be taken on by the other enterprise. The minority participants in each enterprise are bound to provide their share of the investment required (according to the participation shown in Figure 12-2) up to the level required to meet that part of the business which is shared on the one-third/two-thirds basis.

REFERENCE

Zippe, G., J.W. Beams, and A.R. Kuhltan. "The development of Short Bowl Ultracentrifuges," Progress Report No. 1. (ORO-210 (UVA/ORL-2400-58-PR-1)

Chapter 13

Role of the IAEA in Multinational Fuel-Cycle Centers

D.A.V. Fischer

PRELIMINARY STUDY

In late 1974 the IAEA made a preliminary study to ascertain whether economic benefits would result from carrying out reprocessing and waste management at a regional center instead of on a national basis. The IAEA chose Southeast Asia and the Far East as the region of study; it contains a wide diversity of countries with varying commitments to nuclear power.

As will be seen from Table 13-1, three different strategies for reprocessing and waste management were examined, and the costs of each strategy were compared with those of a fuel storage strategy. Table 13-1 assumes a cost index of one (1.0) for the estimated capital investment in a single regional facility and shows the relative costs of other strategies, if followed, in the regions concerned.

Figure 13-1 shows that part of the nuclear power generation costs which is attributable to reprocessing and radioactive waste management. It will be seen from the figure that, aside from transport costs, a country having an installed

Table 13-1. Cost Index Comparison of Fuel Management Alternatives

	Capital Investment	*Total Operating Charges Over 20 Year Period*[b]
Fuel Reprocessing Strategies:[a]		
A Regional Facility	1.0	2.6
Two Regional Facilities	1.1	2.9
Local Facilities	1.5	4.1
Fuel Storage Strategy	0.6	0.9

[a]Includes waste management through solid waste storage.
[b]Not escalated.

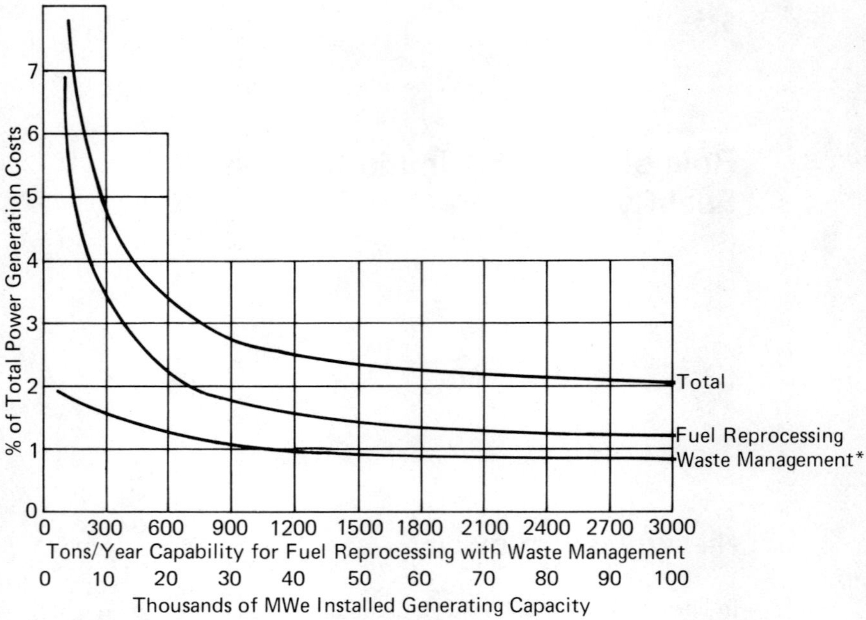

*Waste management costs are through solid waste storage.

Figure 13-1. Index of Fuel Reprocessing and Waste Management Costs for Tributary Nuclear Generating Capacity.

power capacity of 10,000 MWe would pay twice as much a year if it decided to reprocess fuel and handle wastes at a national center as it would if its used a regional center serving a tributary nuclear generating capacity of 50,000 MWe. For a country having less than 10,000 MWe installed capacity, the comparative costs of using a national center would be much higher.

The preliminary study thus showed that there could be significant advantages from regional pooling of reprocessing loads and that a further more detailed study would be justified.

THE REGIONAL CENTER PROJECT

Based on the results of the preliminary study, the regional center project was accordingly launched in September 1975 with the following objects:

1. To develop the methodology for assessing alternative strategies so as to be able to evaluate the advantages and disadvantages of a regional nuclear-fuel-cycle center and of dispersed national or local fuel-cycle facilities.
2. To prepare a report on the methodology, including illustrative examples which could be used by member states.

3. To provide a forum which member states could use to work out alternative strategies.

Scope

The concept of the regional center would be kept flexible to permit various forms of cooperation and to take account of the varying requirements and approaches of the regions of the world.

The project would cover transport, storage, processing and recycling of nuclear fuel from the point where the spent fuel leaves the power reactor through all subsequent steps until recycled fuel in the form of fuel elements is ready for shipment to the reactor. Production of new fuel (that is, nonrecycled fuel) and enrichment would not be included. Management of radioactive wastes would be included.

A center, to be economically viable, will have to be of a certain minimal or optimum size. On the basis of previous studies, it appears that the optimum capacity of the fuel reprocessing facility of such a center would be between 750 and 3000 tonnes of annual throughput of uranium. Hence, the study would estimate on the basis of projections of electrical capacity and of nuclear power growth, the earliest dates at which a given region would need such a center. Subsequently, the study would consider special circumstances in which plants below 750 tonnes might be desirable.

The first step would be to develop a mathematical model that would take into account all the important steps of the fuel cycle and that could be adapted to show the effects of variations within certain limits of each of the major parameters. Figure 13-2 shows the fuel-cycle concept to be used in the study. Further detailed studies would deal with the legal, organizational, financial, environmental, and other questions involved.

Implementation

Four phases are foreseen as shown in Figure 13-3. The first is the preparation of the mathematical model and computer codes. The second is the testing of the model and the code. The third includes a report on the concept, the model, costs, input data, and illustrative studies. The project would also review the experience of other international and regional undertakings to help identify the major legal, organizational, financial, and environmental problems to be dealt with and to develop various approaches to these questions. On completion of the third phase, it will be possible to apply the model to given situations and evaluate specific proposals.

An interim report presented in late 1975 showed that:

a. The region of Central and South America would reach the 750-tonnes-per-year level in about ten years' time.
b. Southeast Asia and the Far East, including Australia, would reach that threshold in 1982.

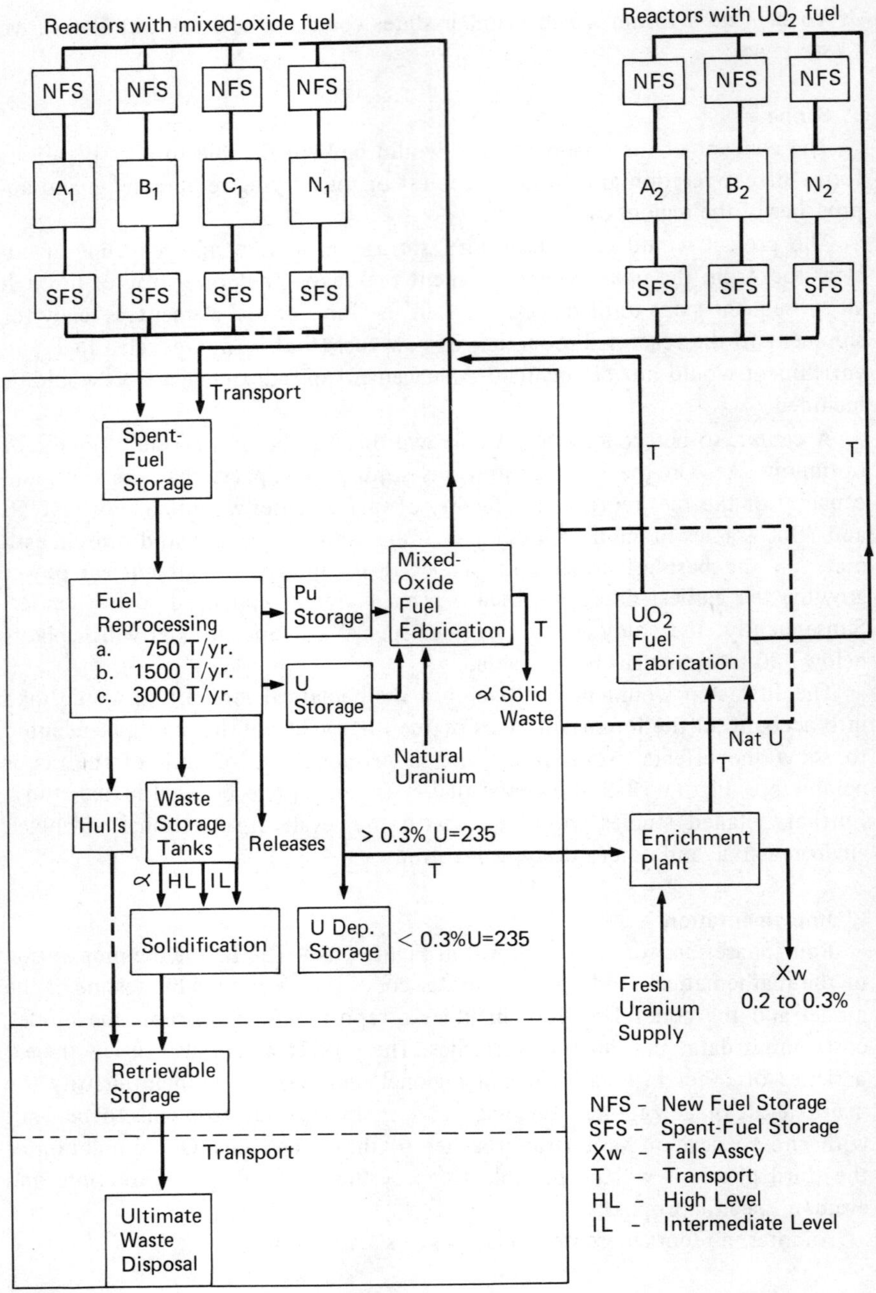

Figure 13-2. Fuel-Cycle Concept.

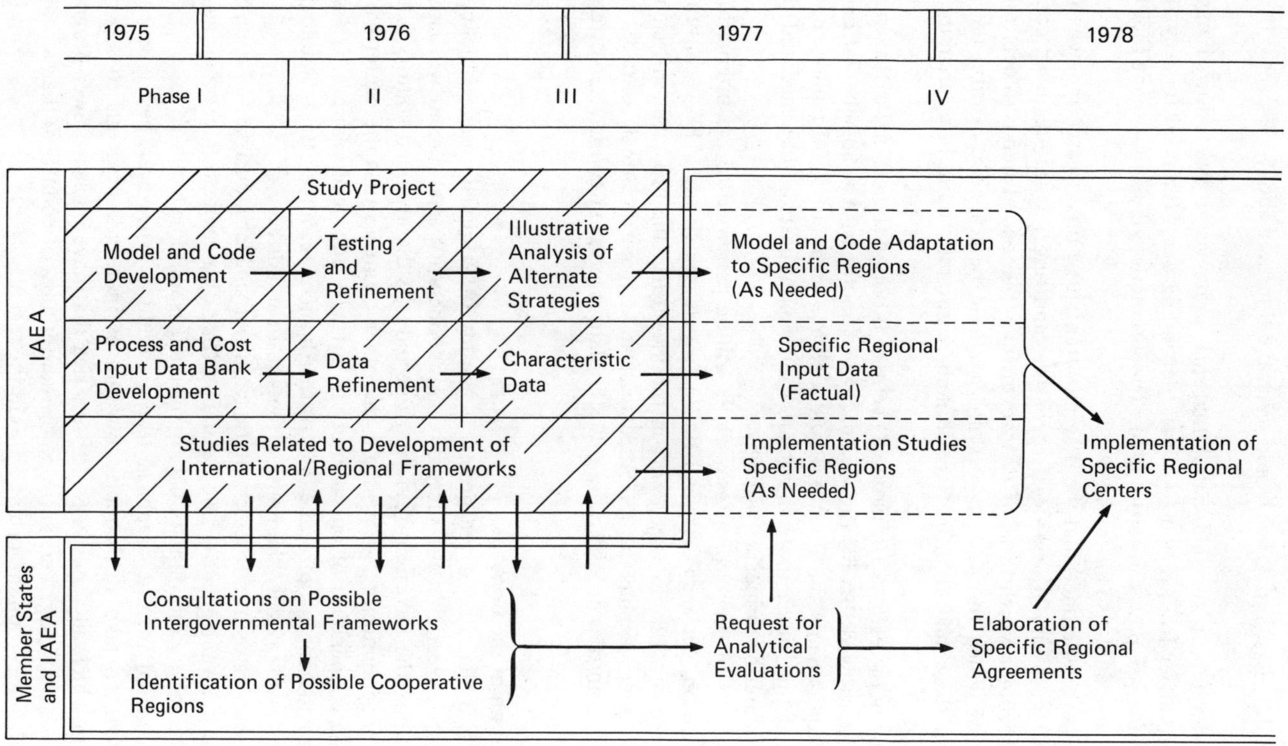

Figure 13–3. Regional Nuclear Fuel-Cycle Centers: Schematic Study Project and Implementation Planning.

c. The Middle East and Africa would reach the threshold about 1990.

The interim report also illustrated that the strategies to be followed might reasonably differ from region to region, depending upon the need for recycling and the type of power reactors used. The alternative of storing spent fuel for several years could also be considered.

The report underlined that the first important question is to ascertain whether it is at all necessary to reprocess spent fuel and under what economic conditions it becomes worthwhile. If it is worthwhile, the second step is to determine the time at which the necessary facilities should be available. In turn, the decisions on these matters affect enrichment requirements and total uranium requirements.

Financing of regional (or national) centers is bound to be a major constraint. The report showed that for the second region (Southeast Asia and the Far East), annual requirements for finance would stabilize at about $400 million, but it is essential that much more information on capital costs be obtained and that these estimates be refined accordingly.

The report on the project as a whole will be presented to the International Conference on Nuclear Power and Its Fuel Cycle, which the IAEA is convening at Salzburg, Austria, in May 1977. In the meantime, the project is sufficiently far advanced to enable the IAEA Secretariat to make informal contacts with groups of countries to ascertain and stimulate interest in cooperation arrangements.

Evolution of the Concept

Although the title of the project is a study of "regional" nuclear-fuel-cycle centers, it is clear that the success of any cooperative arrangement will largely depend upon the inclusion of at least one industrially advanced country which may be geographically remote from the other countries that the center would serve. The cooperation of one or more industrially advanced countries is essential to provide, for instance, the bulk of the financial resources, the technological know-how, and possibly the political dynamism and drive.

In certain cases it is foreseen that a group of smaller industrialized countries may find it advantageous to cooperate in setting up a center to meet their needs. In short, a variety of international arrangements should be explored.

The concept of a geographical region should not, however, be too strongly attenuated. To take an example, it would be more advantageous to set up a project in which all states in South America that have nuclear power programs would take part (together with one or more industrial countries) than a project which left out certain important states in the region.

It may also be desirable to examine further the extent to which it will be necessary to store spent fuel before proceeding with reprocessing. Storage can only be an interim solution, but the advantages and disadvantages of regional and national storage (at reactor sites) require study. Factors to be taken into account include the sensitivity of certain countries to the storage of increasing volumes of spent fuel in their own territories, the timing and the growth curve of the market for plutonium for use in recycle and eventually in breeders, and, of course, the economic and technical problems that long-term spent-fuel storage would pose.

Other questions requiring further study are the extent to which the customers for plutonium from a multinational center may differ from the countries that provide the irradiated fuel. The latter may be more interested in securing supplies of enriched fuel than in a lien on the plutonium produced by the center. Another question requiring study is the extent to which separated plutonium should be stored under international auspices pending its use in recycle or in breeders.

INCENTIVES AND DISINCENTIVES FOR PARTICIPATION

Besides carrying out conceptual studies and developing a methodology for evaluating alternative strategies, the IAEA can obviously play a useful role in bringing together potentially interested countries. It is a truism that the launching of any multinational project depends upon the *political* decisions of the states concerned and this in turn on *their* perception that their national interests are better served by regional cooperation than by nuclear autarky. In short, the major impetus and drive must come from the user countries. The IAEA's own experience has shown that without strong political drive in the countries directly concerned, the most rational schemes for multinational cooperation may come to naught. This, in the author's view, is the most crucial and the most difficult aspect of the project. Let us examine briefly some incentives and disincentives that may operate.

Political

The desire to reduce dependence upon the very limited number of commercial reprocessing services now available may be a strong incentive to the establishment of multinational centers, an incentive enhanced by the fact that such centers would also provide other services or secure their provision—fuel fabrication, waste management and disposal, and possibly enrichment. Another factor that may influence certain countries to support the multinational concept is the strong objection of powerful bodies of public opinion to the construction of reprocessing, waste-handling, and waste-disposal facilities in the territories of those countries.

Economic and Financial

It has been amply demonstrated that multinational projects will enable the participating countries to achieve more easily the economies of scale or supply that would otherwise be denied to them. However, practical experience in this matter is extremely limited, and there is still much uncertainty about the actual capital costs of large reprocessing plants.

In countries where such plants would be built by private enterprise, regulatory requirements and delays may have a very marked influence on costs. Estimates of the investment required in a plant processing 800–1000 tonnes per year vary from $500 million to $1.5 billion. The scale of investment required illustrates the importance of industrial country participation in any project.

Technological

For a variety of technical and regulatory reasons, no large reprocessing plant handling highly irradiated oxide fuel—by far the most common spent fuel—is in full operation at time of writing. Entry into operation of the Barnwell plant and reentry of the NFS plant in the United States have been delayed by regulatory requirements. The Windscale plant in Great Britain is out of action for the time being. The Cap La Hague in France has just begun testing on irradiated fuel.

More R&D is desirable on the reprocessing of highly irradiated oxide fuels. There is, nevertheless, reasonable assurance that technological progress and experience will permit the building and operation of a number of large-scale multinational reprocessing facilities in the time-scale foreseen. Technological considerations are not likely, therefore, to be a major constraint. They may even serve as an incentive by enabling countries participating in a multinational center to acquire access to reprocessing technology through such a center instead of by following the national path.

At this point, it is perhaps desirable to draw a distinction between the plant and the know-how required for the small-scale and uneconomic reprocessing of relatively lightly irradiated fuel from small research reactors on the one hand and, on the other, that required for the operation of a very large commercial center handling highly irradiated fuel from light-water power reactors. The know-how involved in the first case is already rather widespread and the equipment required is fairly easily obtainable. The know-how involved in a large-scale commercial operation is a hard won and commercially valuable commodity, and the plant is relatively intricate and costly.

Plant Safety, Waste Management, and Plant Security

There seems little doubt that compared with a number of smaller and scattered national fuel-cycle facilities a single large fuel-cycle center offers considerable advantages from the points of view of plant safety, waste management and disposal, and plant security. The extent to which these factors will operate in favor of the multinational facility will depend to some extent on the precise location of the facility and the standards required.

Conservation

Considerations of conserving scarce resources (reprocessing irradiated fuel so as to recover the plutonium and uranium instead of discarding the fuel as waste) may be important in the future. It is conceivable, too, that certain groups of countries may be prepared to cooperate in demonstrating the feasibility of the multinational concept so as to set an encouraging example, but these incentives will clearly be secondary to commercial and technological considerations.

THE TIME SCALE

From the time the decision is taken to set up a multinational center, a further ten years or so will be needed to bring the center into operation. The time required for prior negotiations is obviously less easy to estimate, perhaps 3 to 5 years. It is important that such decisions be taken as soon as possible. Otherwise, there is the obvious risk that the countries concerned will not wait for the results of regional initiatives but will proceed along the national path.

First Steps

The IAEA project will provide the necessary mathematical and methodological tools for evaluating all the main strategies and options that are open to the IAEA's member states. The first subsequent step must therefore be to acquaint member states with these tools and to encourage them to apply these tools in the evaluation of their own national programs and problems.

The IAEA Secretariat is taking steps to encourage participation by potentially interested countries in the project so that they may familiarize themselves with the methodologies being developed and the various legal and institutional arrangements that could be used for setting up such centers. The Secretariat is drawing heavily on the experience acquired by various regional and multinational nuclear energy organizations.

The first task is to encourage working parties for groups of countries to estimate the quantity of irradiated fuel that will have to be dealt with by the countries concerned, nationally and in aggregate, until a given deadline, the year 2000, for example. The results of such studies are, of course, in no way binding on the countries taking part, and participation in the studies is without commitment as to what strategy should be followed for dealing with reprocessing and storage needs. However, these studies are in themselves a useful operation from the point of view of the countries concerned.

At a later stage, the countries would consider the alternative strategies for dealing with the spent fuel—that is, long-term storage or early reprocessing. If early reprocessing is the indicated strategy, then three choices must be examined:

1. The use of existing facilities.

2. The establishment of a multinational facility.
3. The establishment of a series of national facilities.

Finally, if the advantages of the multinational approach are persuasive, the IAEA would provide assistance in drawing up suitable institutional and legal arrangements.

It should be stressed that at this stage no country has done more than approach the first task. In certain cases, one of the problems to be solved may be that of integrating reprocessing plants that have already been supplied or committed, into a multi-national framework. It is likely that the main supplying countries could, if they were prepared to do so, play an important role.

SAFEGUARDS AND NONPROLIFERATION

Summing up, the multinational-fuel-cycle-center concept presents a number of major economic, technological, and safety advantages. It should provide a desirable alternative to national self-sufficiency for countries whose chief concern is to find means of satisfying their commercial reprocessing requirements in the next decades.

There may well be other motivations for wishing to acquire reprocessing technology: the desire eventually to minimize dependence upon external supplies and services and, indeed, genuine scientific and technical interest. The multinational fuel-cycle center should also provide an outlet for such scientific, technical, and long-term commercial interests. Thus, for instance, the OECD/NEA Dragon Project in the United Kingdom enabled a number of smaller industrial countries to take part in the development of high-temperature reactor technology without having to set up national facilities of their own. The Eurochemic reprocessing plant served a similar purpose.

A reprocessing plant having a substantial throughput may require continuous inspection. It seems likely, prima facie, that a facility designed and operated under multinational auspices will not only permit the application of more effective safeguards but will also provide substantial economies in safeguards manpower. It has been calculated that a reprocessing plant with a throughput of 100 tonnes per year may require 6 to 8 man-years of inspection; a plant of 1500 tonnes throughput would require 15 man-years. Co-location of other facilities (fabrication and so on) will also reduce safeguards costs.

It must be borne in mind, however, that the participation of a country in a multinational fuel-cycle center will not necessarily prevent that country from proceeding to build or continuing to operate a small reprocessing plant if it is determined to do so. Unlike enrichment technology, reprocessing technology of the kind in question has long since been made public (since 1958, to be precise). As pointed out previously, the construction of a small plant to reprocess lightly irradiated natural uranium from a research reactor is within the reach of several countries having reasonably advanced industrial infrastructure. In fact, it is easier to build a small reprocessing plant than a power reactor. An embargo on the export of reprocessing technology and equipment would make it more difficult, time-consuming, and costly for the affected countries to construct such facilities, but it would not put them out of their reach.

Thus, while international fuel-cycle centers may provide a powerful disincentive to the proliferation of small and uneconomic reprocessing plants, they do not, ipso facto, preclude such proliferation. The multinational fuel-cycle center is one of several concepts that would reinforce our efforts to prevent the further spread of nuclear weapons, but I believe that it is inevitable in the long term that more and more countries will be able, if they so wish, to make use of advanced nuclear technologies. It is therefore of cardinal importance that the legal regime created by the NPT and the international safeguards that accompany it, or comparable international safeguards on all national nuclear activities, should be as widespread as possible and in particular should apply in those countries that stand on the threshhold of advanced nuclear technology.

Chapter 14

Nuclear Futures for Sale: Issues Raised by the West German-Brazilian Nuclear Agreement

William W. Lowrance

When foreign ministers Hans Dietrich Genscher and Antonio Azerada da Silveira signed their countries' fifteen-year Agreement of Cooperation in the Peaceful Use of Nuclear Energy in Bonn on 27 June 1975, the event made headlines all over the world. The Brazilian press jubilantly proclaimed the "dawn of the nuclear age in Brazil" and the economic independence it would bring. Welcoming the "12,000,000,000 Mark für Deutschland," West Germans hailed the "breaching of the U.S. monopoly" that this "nuclear deal of the century" represented. Chang Te-Chung, Ambassador to Brazil from the People's Republic of China, expressed "great satisfaction" over the signing, "because the atom ought to serve all humanity and not be the monopoly of a few powers." Argentine military strategists began to reappraise the balance of nuclear weapons potential in Latin America. *Pravda* commentator Yuri Kharlanov noted that "one can hardly oppose the use of new forms of energy, including atomic energy. However, it is essential that the provisions of the Nuclear Non-Proliferation Treaty (NPT) be strictly observed."

Calling the deal "nuclear madness," the *New York Times* editorialized that the move was a "tragedy for West Germany as well as mankind as a whole." In words so sharp that they penetrated ministerial chambers around the globe, Senator John O. Pastore, chairman of the Joint Committee on Atomic Energy, angrily complained that Bonn was "creating a likely peril in our backyard, while we are heavily engaged in Germany's backyard to defend them against likely peril"–to which the Brazilian retorted that they were not anybody's backyard, and the Germans, pointing out that this agreement was the victory parchment of intense commercial competition, simply scoffed, "Sour grapes." Thus it was no exaggeration when the *Neue Rhein-Zeitung* characterized the agreement as "the most controversial foreign deal ever struck by West German industry."

Regardless of how they may come to be judged by history, the events that elicited the above responses marked the crossing of major thresholds both in technology transfer and in international politics. The present paper will review those events, discuss their unfolding and their present implications, and then draw a more general description of the political territory in which they are long bound to remain prominent landmarks.

By way of overview, we will note that the agreement provided for the largest industrial nuclear sale ever transacted, its total value exceeding DM 12 billion ($US 5 billion). More importantly, it was the first time a complete, self-sufficient nuclear-fuel-cycle "package sale" had ever been made between nations. Its most controversial feature was its inclusion of the two "sensitive technologies" for enriching uranium and reprocessing spent fuel, both of which, if sufficient effort is expended, can be applied to the making of nuclear-weapons-grade fissile material. The supplier nation is a signatory of the Nuclear Non-Proliferation Treaty and has long been a supporting strand in the web of nations committed to retarding the spread of nuclear weapons; the recipient nation, a nuclear newcomer, has refused to sign that treaty and in the past has made few such commitments.

Pervading the present international debate over these events are issues of the developing nations' rights of access to the political, military, technical, and economic wealth that nuclear technology holds, on the one hand, and on the other, management of the export competition in such a way that the technology can be shared without compromising international stability.

For those who worry that the spread of nuclear power technology will also spread nuclear weaponry, the essence of the issue resides, like the arms-controllers' classical "dove's dilemma," in what we might call the "nonproliferator's quandary": Should supplier nations refuse to export critical and sensitive technologies and take other embargoing actions, thereby buying time and prolonging hegemonical control but risking offending the developing nations seriously and inciting indigenous development of the capabilities? Or should they acquiesce in the foreign import-pull and domestic export-push and go ahead and supply the technologies, but negotiate uncompromisingly in exchange for sanctionable promises of restraint on use of the technology? What follows is an exposition of the subject events as they relate to that quandary.

BACKGROUND TO THE AGREEMENT

That Brazil would turn to nuclear power was inevitable. It had a population exploding from 41 million in 1940 to 105 million in 1974, bulldozers being sent out over half a continent, a storied land slightly larger than the continental United States. It boasted an economy with a gross national product of $US 47 billion, but its exports still were exceeded by imports. Its mushrooming cities, new transportation systems, and proud young industries were growing rapidly to meet consumer demands and to become able to convert mineral and other commodities into as refined a form as possible before releasing them onto the export market. A commensurate energy economy quadrupled from 1940, when about 77 percent of Brazil's energy was still derived from wood, charcoal, and sugarcane bagasse, to the present diversity of fuels. And all of this was happening in a part of the world where the kind of prestige that nuclear technology connotes is high political capital.

Of Brazil's total energy consumption in 1972, 45 percent came from petroleum; 30 percent was produced from cellulosic materials; 20 percent was hydro; 4 percent came from coal; a negligible amount came from natural gas; and the order for the first nuclear plant had just been let, to Westinghouse. Domestic oil is not estimated to be extractable at present levels for more than about fifteen years, and little natural gas is available for fuel; two-thirds of the oil is imported. There are reserves of shale oil, but these have not been developed, and offshore oil is an uncertain factor. Hydropower potential, although always cited correctly as being among the world's largest, is difficult to tap because much of the potential lies at the headwaters of the Amazon and its feeders in the north, far removed from the industrial centers of the southeast. For instance, the promising Tapajós and Xingu basins are more than 1500 rugged kilometers north of São Paulo; the powerline construction and maintenance costs and the losses in long-distance electrical transmission are discouraging. Currently 15 percent of the hydropotential is harnessed, generating 95 percent of the total national electric output of 16 gigawatts.

Nuclear power had been in the long-range plans for many years. Maturing slowly through the late sixties and early seventies, being set back by economic misfortunes and changes of government, Brazil's nuclear power policy eventually came to have several critical features. More than its neighbor and rival, Argentina, Brazil tended to favor substantial importation of the required technology over developing it indigenously. Brazil chose to adopt light-water reactors using enriched uranium, the principal systems used in Europe and the United States, instead of heavy-water reactors fueled by natural (unenriched) uranium, as exported principally by Canada. Argentina was exploring both reactor types, but was favoring the latter. Progress followed the usual course of signing various binational agreements, building research reactors, training specialists, and shopping around for the first major purchases.

The three most promising sources of technology were West Germany, France, and the United States. Independently all of these aided Brazil in charting its energy future. West Germany was able to take advantage of the strong industrial connections it had long had with Brazil. Lufthansa had assisted the flag air-carrier Varig for years; the great Ruhr steel tubemaking firm Mannesmann had operated an important subsidiary at Belo Horizante; and the Brazilian Volkswagen factories had been among the most successful anywhere. The nuclear ties can be traced as far back as the bizarre 1953 episode of the attempted secret shipment from Germany to Brazil of three uranium-enrichment centrifuges, aborted when James B. Conant, U.S. High Commissioner for Occupied Germany, ordered the already crated machines impounded because their export would be a violation of the conventions of the close of the war; the mystique that collaboration imparted still lingers as an overtone in external discussions of the two countries' relations. But since that time Brazil and Germany had carried out several joint ventures successfully. The United States too had undertaken many efforts with Brazil; its business investment had been heavy. France had been less involved in Brazil historically, and seemed interested principally in joint uranium ore ventures.

Brazil took bids for its first reactor in 1967. Westinghouse won, and began building a 600 megawatt plant, Angra-1, on the coast above São Paulo at Angra dos Reis (startup scheduled for 1978). As part of the contract, the first charge of fuel and two reloadings were guaranteed by Westinghouse.

The nuclear discussions continued over the next few years, accelerating in 1973 as the oil crisis made its punishing effects felt. January 1974 brought a restructuring of the nuclear establishment; responsibility for nuclear power development was delegated to a state utility, Empresas Nucleares Brasileiras, or Nuclebrás. Nuclebrás picked up the negotiations with the supplier nations and vigorously pursued the keystone purchase that forms the subject of this paper. Shuttling from their capitals and corporate headquarters to Brazil, the West Germans and the Americans fought for the contracts.

For a while Westinghouse thought it had secured the sale of Brazil's next several reactors, but that position was eroded by two factors. First, in mid-1974, because of uncertainties about the capacity of the U.S. government's uranium-enrichment plants (such as that at Oak Ridge) to supply fuel for future reactors, the Atomic Energy Commission shifted some long-term fuel guarantees to "conditional" status and returned the deposit payments on them. The contracts for fueling Brazil's second and third (as yet nonexistent) reactors were affected. Coming so soon after the previous year's oil restrictions, this threat of fuel cutoff was understandably disturbing to the Brazilians. And second, in late 1974 Westinghouse began to learn that the West Germans were making aggressively competitive offers and, supposedly at Brazilian insistence, were considering selling not only power reactors but also the sensitive technologies of enrichment and reprocessing.

This became the point of angry contention. The facilities for enrichment are used to refine the fissile form of uranium out of the ore up to the three percent concentration required for LWR fuel; with an appropriate effort, though certainly that effort would require making substantial modifications to the plant, it appears possible to refine that uranium up to the high concentrations usable for nuclear explosives. Reprocessing facilities are used to receive hot, radioactive spent fuel discharged from reactors and recover unburned fuel to be recycled through the reactors. Unfortunately, this process also allows for extracting plutonium, an unavoidable by-product of the uranium fission that goes on in reactors, and plutonium makes at least as good a bomb as highly enriched uranium does.

It had always been the policy of the United States to refuse to sell those technologies. Reactors were fine, but because of their weapons implications, enrichment and reprocessing were never exported. When the United States got signals that the Brazilians were trying to buy the entire fuel-cycle capability, and that the West Germans were apparently willing to include the sensitive technologies in order to clinch the sale, the discussions became highly charged. Then Westinghouse learned that it was out of the bidding.

For the next months the negotiations remained the province of the few professionals involved on the various sides, but they were not kept secret. Brazil concluded its negotiations with West Germany on 12 February 1975. A week later U.S. Ambassador Martin Hillenbrand was informed in Bonn, and at the same time in Washington Ambassador Berndt von Staden briefed Arms Control and Disarmament Agency director Fred Iklé. A news item about the decision appeared in 20 February issue of the widely circulated American industrial magazine *Nucleonics Week.* On 7 April a four-man delegation from the State Department and the arms control agency traveled to Bonn to make representations; obviously the team's protests over the sensitive technologies did not stop the sale, but it is said that they did induce the Germans to take a stricter approach to the safeguards. There were other discussions between the governments, but apparently not at the highest levels. On all sides the decisions were handled mostly by industrial and ministerial officials. The Bundestag did not become involved, nor did the Brazilian or American Congress. Because of the delicacy of the discussions a moratorium on articles about the deal was imposed on the Brazilian press for three months of the spring. No major stories appeared in either German or U.S. news publications until late May. Pursuant to its obligations under Article 103 of the Euratom Treaty, the European Commission reviewed the prospective export and, on 5 June, sent word from Brussels that it had no objection. Finally, on 27 June, with much diplomatic pageantry, the bilateral agreement was signed.

THE AGREEMENT

An umbrella document, the agreement opened the way for a large number of commercial contracts between West German firms and Brazilian firms and utilities, and it obliged the signers to establish safeguards and protect the nuclear goods from such hazards as theft and sabotage. Because of the Nuclear Non-Proliferation Treaty's restrictions binding the Bundesrepublik, the agreement could go into force only after both nations had signed and ratified an appropriate trilateral safeguards accord with the International Atomic Energy Agency (as it turned out, this would be accomplished on 24 February 1976). The agreement included a declaration "affirming the principle of nonproliferation of nuclear weapons," promises that the materials would not be used to make "nuclear weapons or other nuclear explosives", and a requirement that all re-exports be subject to approval by the German Federal Republic.

THE COMMERCIAL ARRANGEMENTS

The many contracts being let under the pact are intended eventually to provide Brazil with an autonomous nuclear industry. The business arrangements are complex. Most of them begin with binational financing and development and shift over the period of the agreement to sole Brazilian ownership.

Uranium prospecting, mining, and conversion will begin immediately, with Nuclebrás (51 percent ownership) working with Urangesellschaft (49 percent). Appraisals of the uranium reserves vary widely. West Germany depends on imported uranium, since it has little native ore and is rapidly building its own nuclear plants, so this part of the trade is cited as being important for Germany.

Two power reactors of the 1250 MWe Biblis pressurized-water type are to be built by a combine of Kraftwerk Union AG (75 percent) and Nuclebrás (25 percent). The first plant, now being started on, will be manufactured almost completely in Germany, assembled on the Angra River site, and transferred on a turnkey basis. The start on the second is scheduled for 1977. Each will take about six years to build. The more than DM 4 billion financing for these two plants has been provided by a consortium of banks headed by the Dresdner Bank and aided by the Kreditanstalt für Wiederaufbau. Down payment is 10 percent and the basic loan rate to Brazil is a low 7.25 percent, to be repaid over a period of twelve years after the plants begin producing electricity. The early announcements described an eight-reactor program valued at around DM 12 billion, to be completed within the fifteen-year period of the accord; so far, only the orders for the first two reactors are firm.

A heavy-components facility is to be built in Brazil by a consortium of Nuclebrás, Kraftwerk Union AG, Gutehoffnungshutte Sterkrage AG, and the Austrian firm Vereinigte Österreichische Eisen-und-Stahlwerke (Voest).

A pilot fuel-element fabrication plant is being planned by Nuclebrás (70 percent) with Reaktor-Brennelement-Union GmbH (in which General Electric is a partner) and Kraftwerk Union.

The controversial uranium-enrichment facility is to be based on the as-yet commercially unproven "jet nozzle" process. A demonstration plant is to be developed in Brazil over the next five years by Nuclebrás (75 percent), Steag AG (15 percent), and Interatom (10 percent), with expansion to commercial scale to follow when feasibility has been confirmed. This process has been worked on for years by Dr. Erwin-Willy Becker, first at the Karlsruhe national nuclear research center and then, since 1970, at Steag in Essen. This plant is one of the more inexplicable aspects of the whole affair. Although the Becker process has been worked on for years and has been supported by substantial government investment, the Bundesrepublik has never been willing to give it the high priority it has granted the process based on centrifuges. The nozzle process has the intrinsic disadvantage that it consumes far more electricity than the alternative techniques do; criticism on this point has been answered with the reply that in Brazil the plant will be constructed near some large dam up in the mountains, utilize surplus electricity that would otherwise be wasted, and in essence "encapsulate" the hydropower in the form of fuel pellets from which the energy can later be released by reactors in the industrial regions. The enrichment site has not yet been announced. For Brazil, investing so heavily in this novel technique, which has never been used on a commercial scale anywhere, even in Germany, surely represents a high risk. Transferring such an untried and exceedingly complex technology contradicts all that has been so painfully learned about transferring high technologies to industrializing nations.

The reprocessing plant, a small one (10 kilograms fuel per day) is to be built by Nuclebrás with a consortium of Kernbrennstoff-Wiederaufbereitung GmbH (KEWA) and Friedrich Uhde GmbH.

A number of other contracts are included in the package, and these will lead to subcontracts. It will take years to let all the jobs, arrange financing, and actually build the facilities. How this will all turn out is difficult to predict. No doubt the reliable Kraftwerk Union reactors will serve well. And others of the technologies are beyond question. But troublesome aspects are reflected, for instance, in the skeptical comment now in print, with slight variation, in several languages: "The Germans have sold an enrichment process that doesn't work, to enrich Brazilian uranium that doesn't exist." It remains to be seen whether the overall plan will lead to the sophisticated indigenous Brazilian industry envisioned. And beyond all this there are the issues of the proliferation of nuclear weapons.

OBJECTIONS BY THE UNITED STATES

The American intervention efforts—the "Querschüsse," or shots across the bow, as they were referred to in the German press—took the following expression (assembled here as a composite of various statements):

> It is entirely reasonable for Brazil to build reactors. And that implies the necessity to train a technical infrastructure and build an industry. But an enrichment plant makes much less commercial sense, being very expensive and, with the Becker process, unproven. The enrichment venture is irrelevant to building a basic electric power industry and is wasteful of Brazil's development resources. This small plant cannot ensure uranium fuel independence, and it is hard to imagine even token export of en-

riched uranium before many years have passed. There is no question that, given sufficient technological modification, *any* kind of enrichment plant can be adapted to making bomb-grade uranium.

France has always refused to sell its enrichment processes; Britain has refused to sell its processes; the United States has refused to sell its processes; and Urenco (the tripartite centrifuge enrichment venture involving the German Federal Republic, the United Kingdom, and the Netherlands) has refused to sell its process. Thus the sale to Brazil erases the line of restraint that the Alliance nations, recognizing that they are the tragic custodians of a double-edged technology, have always felt obliged to maintain. An enrichment plant is not just a simple item of commerce. As to the unchallengeable argument that the Brazilians have a need for a secure supply of uranium fuel, a much more immediate and reliable way to fulfill that need would be to consign Brazil a certain guaranteed amount of the Bundesrepublik's own allotment from Urenco, or help make some similar firm supply arrangement.

The reprocessing plans make even less sense. No recycling operation has ever been a standard part of any civilian nuclear power cycle. Small facilities have been built in a number of countries, but they have never proven commercially essential or profitable. For example, the United States, with tens of reactors in full operation, does not yet find any overall economic advantage in trying to reprocess spent fuel (although as the industry grows and reactor technologies evolve, this may change). Reprocessing plants have indeed been operated for decades in the United States and elsewhere—for making bombs. This has been the source of all of the plutonium, and thus of most of the atomic bombs, in the arsenals of the nuclear weapons states. It was also the technology the Indians used to extract the necessary few kilograms of plutonium from fuel discharged from a small Canadian reactor and assemble the bomb that rocked the Rajasthan desert in 1974. The reprocessing plant is the most troubling aspect of the entire transaction.

It must be recognized that the political context for the trade is not at all reassuring. Although the situation appears to be relatively stable at the moment, Brazil's political history has been marked by recurring upheavals of government, internal repression, terrorism, and sabre-rattling toward neighbors. Brazil has steadfastly refused to subscribe to the Nonproliferation Treaty, and it has never been more than a reluctant participant in discussions of nuclear restraint.

THE BUNDESREPUBLIK'S RESPONSE

The West Germans make the following counterarguments (again, assembled here as a composite).

The day of the industrial nations' hegemony is over (perhaps particularly that of the United States). The developing countries cannot and should not be denied; they will no longer tolerate, nor do they have to tolerate, the kind of discrimination they have had to suffer in the past. Their wealth and political power, which have long been underestimated, are now going to be felt. With elegant new techniques coming out of the laboratories, the technological hurdles that once kept back all but the strongest have gotten lower, while the contenders have all gained technical strength. If the developed nations refused to supply the technology, the developing ones will simply make it themselves; that will take just a little more time, it will make those countries much more capable in the long run, and it will lead to just the kind of alienation that we must try to avoid. Another change is that the suppliers, once only a fastidious few, have now become many. So if the Federal Republic declines to sell, the buyers will simply turn to France, Japan, or somewhere else. When one recognizes that all of these critical changes–the weakening of traditional hegemonies, the lowering of the technological barriers, and the diversification of sources of supply–have removed the long-term technical obstacles to nuclear weapons capability, and when one then realizes that the only available controls on "going nuclear" are political ones, the only strategy that makes sense is to work to integrate the industrializing nations firmly into the existing web of power. This integration entails transferring technology, but exchanging that gesture for good will and promises of restraint from the recipients. In adopting this approach Germany is simply following the provisions of the Non-Proliferation Treaty's Article 4, which binds subscribers to facilitate "the fullest possible exchange of equipment, materials and scientific and technological information for the peaceful uses of nuclear energy . . . with due consideration for the needs of the developing areas of the world."

Looking at the present case, we note that Brazil's gross national product is the tenth largest in the world and has grown impressively for the past five years. Brazil has been training a cadre of capable nuclear experts. If Brazil had chosen to take the route the Indians did, nothing could have stopped it; and in fact, since Brazil's desire to develop nuclear explosives for massive landmoving projects and so on had long been propagandized, a course like that taken by the Indians would not have been surprising. Instructive analogy can also be seen in Argentina, where an impressive nuclear capability has been built up by a combination of diligent internal development and transfers from Canada, France, West Germany, and other countries. What is novel and crucial about the arrangements made with Brazil are the safeguards, which are probably the most stringent in the world. By this strategy, then, this strong young country which would otherwise have made its own way aloof from the restraints of international safeguards, inspections, and formal declarations of intent, has now been brought into the matrix of all of those restraints.

THE SAFEGUARDS

The safeguards agreement is indeed as strict as any international safeguards accord now in effect. Basically, the arrangement places all special nuclear information, materials, and equipment transferred between the two countries under the regulations of the IAEA's standard safeguards system INFCIRC/66/Rev.2, which provides for inspection of facilities and for inventorying of material on site and in transfer. The nations "undertake that none of the covered items shall be used for the manufacture of any nuclear weapon or to further any other military purpose or for the manufacture of any other explosive device" (Article 2); this provision covers the "peaceful nuclear explosive" possibility. Sensitive goods may be transferred to third parties only if those parties subscribe to safeguards approved by the agency (Art. 10.2). (The bilateral agreement itself gave both countries a veto right on transfers to third parties.) Subsequent generations of special fissionable material are covered. Nuclear material substituted for other nuclear material is in no way exempt (Art. 7); this takes care of the Indian ploy of allegedly substituting Indian uranium for Canadian uranium in Canadian supplied and safeguarded reactors, carrying the material through to the bomb, then claiming that no safeguards agreements were broken. Transferred information itself is to be inventoried and safeguarded:

> Any nuclear facility or specified equipment designed, constructed or operated, within a period of twenty years after the communication made to the Agency . . . in the State to which relevant technological information has been transferred shall be deemed to be designed, constructed or operated on the basis of or by the use of transferred relevant technological information if its design, construction or operation is based on the same or essentially the same physical or chemical process or processes as those originally described to the Agency (Art. 3.2).

The presumption established by this provision is designed to ensure that all similar facilities constructed during other 20-year period will be under safeguards, without the necessity of proof that transferred technology was actually used. And if, after the safeguards agreement has ceased to be in force, a facility is "designed, constructed or operated in either State on the basis of or by the use of relevant technological information transferred from the other," the safeguards agreement will "forthwith be reinstated" (Art. 28.2). As of 24 February 1976 this agreement was placed in force, whereupon commercial execution began.

The safeguards system is one of accountancy. Its only power is that of detection. Its only force is deterrence by the possibility of discovery. If the inspections or audits find any discrepancy, such as a shortcount of special fissionable material, it is provided that a series of notifications will be made: the offending country is to be given a chance to rectify the situation; failing satisfaction, notice is to be given to the IAEA Board of Governors; ultimate appeal is to the Security Council of the United Nations.

The internal problems of the IAEA safeguards system are not trivial. They have to do with imprecisions in the basic accounting procedure, with the vulnerability of this so-called "technical" process to politicization, with the system's relatively long response time, and with the overwhelming of the agency's limited capabilities by the rapid growth of the worldwide nuclear enterprise. And in the case of the Brazilian safeguards, despite the presumption discussed above, considerable room will remain for quibbling over the question of whether a particular bit of information was used for building a facility, or over how closely a new piece of equipment resembles a progenitor. And if, some years hence—perhaps under a different government—the bilateral agreement is terminated and a new facility is built, for which, by the presumptive safeguards provision, safeguards should "forthwith be reinstated," how is that reinstatement to be enforced? The system deters, but is helpless to prevent absolutely, such actions as clandestine or overt replication of facilities, staged "terrorist" theft of material by the government itself, or transfer of secret information to collaborators. Thus, legalistic reassurances notwithstanding, technical safeguards alone are not very powerful. Obviously agreements can be abrogated, doors closed against inspection, and facilities expropriated.

However, in the present case the safeguards strategy has secured major commitments from a country that had in the past made no such concessions. Although Brazil still has not subscribed to the Non-Proliferation Treaty, for the programs developed with Germany it has now in effect agreed to the spirit of many of that treaty's provisions. It has agreed to waive certain sovereign perquisites and open its facilities to the IAEA inspectors. Further, it has made formal public renunciation of all intention to use the transferred information, materials, and facilities to make nuclear explosive devices. Therefore, within the period of the agreement Brazil cannot make any major steps toward a bomb without domestic and international detection, discussion, and challenge.

IMPLICATIONS FOR THE IMPORTER AND THE SUPPLIER

For Brazil, the purchase will provide several badly needed power plants, at a cost competitive with fossil-fueled alternatives. It will commit Brazil to developing a nuclear industry and infrastructure, the long-range social and economic consequences of which are unpredictable but which may eventually prove troublesome to the country. It will probably have the effect of postponing exploitation of the country's wealth of native solar and hydro energy, and, except for opening an alternative to oil, it can hardly increase Brazil's energy independence. The drain on capital will be heavy. Brazil's external debt, which had risen to $US 22 billion by the end of 1975, will be increased. The purchase will keep Brazil dependent on imported technology for a long time. In other industries there have been complaints that foreign "cooperative" deals have failed to improve native capabilities; it is said that despite long manufacturing activity by such firms as Verolme and Ishikawajima in the port cities, Brazilians themselves are not becoming shipbuilders as expected. Much of the initial nuclear hardware will really be European made; it is too early to know how efficiently indigenous Brazilian competence will be induced. At present it is by no means certain whether Angra-1 will be joined by more Westinghouse reactors, whether Angras-2 & 3 will be joined by the agreed-upon six Kraftwerk reactors by the mid-1990s, or whether, as Brazilian electric utility chiefs have so enthusiastically projected, the country will build sixty-three reactors producing 61 gigawatts by the end of the century. The latter seems unlikely. If the uranium deposits turn out to be major ones, ore export will of course bring in welcome revenue; so far, evidence on the reserves is not impressive. Commercial enrichment of uranium in Brazil is not likely any time soon; Brazil will certainly have to import its fuel.

Brazil is already opening new avenues of cooperation. It speaks of leapfrogging to a breeder economy; it is negotiating for a high-temperature-gas thorium breeder reactor from the United States, and it has recently entered into an agreement with France (called "Cobra", for "cooperation Brazil") to develop a Brazilian counterpart to the Phénix. All this nuclear activity may bring a certain prestige to Brazil with its status-conscious neighbors, and it may strengthen the internal political position of President Geisel. And for the long term it will keep the options of nuclear power and nuclear weapons open.

For West Germany, the sale will bring at least temporary relief to a giant industry burdened by ties to nonprofitable allied divisions and suffering from an almost 30 percent overcapacity. Optimistic projections of the domestic market made in earlier years have led to overdevelopment of the industry. Kraftwerk Union, the larger German reactor manufacturer (the other is Babcock-Brown Boveri), says it is geared up to start about six reactors a year, on a six-year construction cycle; about four orders came in last year from domestic utilities. So the salesmen are being sent elsewhere, looking for countries that have a reasonably sophisticated technical infrastructure–and a stable enough government and enough fluid capital to make payments. Kraftwerk Union is the nuclear subsidiary of the giant electrical firms AEG-Telefunken (50 percent) and Siemens (50 percent). AEG has suffered perennial technical problems with its boiling-water reactors. It is still struggling to get its station at Würgassen on the Weser started up; the facility has had to be shut down three times in as many years–costing AEG as much as DM 210 million on a plant selling for DM 335 million–which has meant an expense, for modifications to all of the plants of that design, based on the Würgassen experience, of up to DM 1200 million. The financial troubles have been so great that AEG has long considered selling its share of Kraftwerk Union, but at present the firms are still locked in partnership. The company would not have survived without massive capital infusions from a banking consortium led by the Dresdner Bank (the same bank, surely not coincidentally, that has underwritten the Brazilian purchases).

Embarrassingly for the government's energy program, Germany has developed a surplus of coal and even some surplus of oil. The piling up of some 20 million tons of unsold anthracite at the pitheads along the Saar and the Ruhr has brought rumors of pit closings, and oil refineries have been cut back to 60 percent throughput. For various reasons the demand for electricity has been dropping. These "difficulties" surely will resolve themselves in time, but in the interim they are contributing heavily to the reactor vendors' headaches. The package sale to Brazil will help both industry and Mr. Schmidt's Social Democratic Party. It will bring subsidization for Dr. Becker's nozzle uranium enrichment work, and it will provide work for the German partner in the reprocessing consortium. Other, conventional enterprises will follow across this trade beachhead into Brazil. All of this commerce will bring much-sought export revenue to the Bundesrepublik.

SOME CRITICISMS, ASSERTIONS, AND CONCLUSIONS

Although some have high merit, many proclaimed "justifications" for including the sensitive technologies in the Brazilian deal have been specious. The simple incontrovertible fact is that the German Federal Republic placed these technologies on the bargaining table in order to clinch the entire sale. This fact remains, regardless of the chronology of events, requests by the Brazilians, or the actions of commercial vendors. Since the beginning of the Nuclear Age the Alliance nations, West Germany especially, have known the American government policy on nuclear exports; there has never been any ambiguity over whether the United States approved of export of enrichment and reprocessing technologies to the Third World.

As recently as 17 March 1975 the West German Foreign Office (responding to a flawed but nonetheless provocative article in the 15 March *Spiegel*) was still arguing that "there never existed a consultation gap" between its government and that of the United States. This glosses over the situation. The issue is at what levels such consultation was undertaken, and with what timing. There is no evidence in the public record to indicate that the responsible officials of the Bundesrepublik consulted their American counterparts until the negotiations with Brazil had reached such an advanced stage that the complicated commercial plans had been worked out and the countries had even exchanged letters of intent. Only at this late moment, on 19 February 1975, did Bonn inform U.S. Ambassador Hillenbrand. There certainly was no joint German-American appraisal before the sensitive technologies were first tendered to the Brazilians. The economic and political momentum the deal was allowed to gather by late spring left little chance that events could be deflected by the Congress, the newsmedia, or by the four-man technical team Washington sent to Bonn in April. On matters as grave as these the preceding can hardly be considered full diplomatic consultation.

Another objectionable aspect of the German position has been the excessive reliance on legalistic arguments. One of these has been that because Article 4 of the NPT requires that signatories promote the peaceful uses of atomic energy, the Bundesrepublik was inescapably obliged to fill Brazil's requests for the sensitive technologies. An ancillary argument has been that the very inclusion of enrichment and reprocessing technologies in the IAEA safeguards "trigger list" (INFCIRC 209, which defines the technologies that should be safeguarded) implies that they should be traded no differently from other routine items. To the first we respond that NPT Article 4 calls for furtherance of the "peaceful uses" of the atom only. To the second we need only point out that a requirement reading, "if this sensitive technology is ever exported, it must be placed under safeguards" does not in any way imply either a license or an imperative to export that technology. An export policy based on literal deference to treaties, under minimalist interpretation, would not be in keeping with the rest of West German's complex, subtle, dynamic foreign policy.

An argument that has proven misleading, particularly to the uninformed, has been the one that says that "no military use can be made of these facilities —because they are under international safeguards." Reassuring repetition of the word "safeguards" notwithstanding, technical safeguards are no more than procedures for conducting moderately close inventories and inspections, with their only power being deterrence by threat of detection. Safeguards are indeed essential. But they are by no means such a deterrent to rash action that all the other factors of international politics can be ignored. Far from it. In fact, in the extreme, the glib assurances of safeguards may simply serve as a Trojan horse allowing dangerous technologies to be conveyed under a guise of benignity. Safeguards agreements can be abrogated on short notice. Extremely serious acts can be accomplished before the United Nations Security Council can take action, and there is not much in the council's history to inspire confidence in its powers of enforcement. Sanctions, meaning political and economic punishments, are often referred to, but, again regrettably, history shows that these have always proven easier to talk about than to enforce. Safeguards are crucial. Safeguards should be insisted upon. But although safeguards are in every way necessary they are by no means sufficient to ensure against abuse of transferred technology over the long run. Similarly, treaties do have value, and the NPT has several important proliferation-retarding provisions; breaking those promises would constitute a major international offense and could not be done except at considerable political cost. A treaty, though, is still just a treaty.

One more unacceptable argument is that the Third World should not be discriminated against, that it is on its way to achieving its destiny, and that no technology should be withheld from it. There is nothing compelling about this argument. All political relations are discriminatory. In negotiating for import commodities, in the exchange of currencies, in securing its defense, the Bundesrepublik is selectively and powerfully discriminatory, as all nations are. Because among the items of commerce the awesome atom is unique, its control demands special policies, even if they are seen as discriminatory in fact. Few have agreed, for instance, with the Bundestag member who argued recently that "we cannot refuse to sell to Koeberg (South Africa) just because we don't like the government there," and nobody believes that Libya is equivalent to Japan as a recipient of a reprocessing plant. Nuclear technology does have to be shared, but carefully. Very carefully. As one aspect of this, introduction of the high technologies by stages makes sense, with transfer of the more marginal, experimental, and sensitive technologies made contingent upon the recipient's demonstrating impeccable practices and unquestionable good intentions with its first reactors and other basic technologies.

It bears repeating here, since the reprocessing plant is probably the most highly contentious part of the sale, that neither the Brazilians nor the West Germans have presented solid justifications of the commercial necessity for this technology. The nuclear nations do not themselves routinely recycle fuel, so why should a country just having its first reactor built be expected to do it?

West German spokesmen, along with others, have expressed several insights that are unchallengeable and strongly deserving of attention: that the technological barriers are coming down, that the spread of nuclear power is inevitable, and that the developed nations must drastically revise their views of the industrializing countries. But for them to say that the United States' policy was not clear, that early consultation was carried out at high levels of government, that the Nonproliferation Treaty left Germany no choice but to sell Brazil the sensitive technologies, that submission to safeguards is sufficient grounds for exporting reprocessing plants regardless of other political factors, that the Third World should not be denied anything at all, that enrichment and reprocessing facilities are essential to Brazil's civilian power development—these arguments are entirely specious. As are the even more lame rejoinders that the United States' hands are not clean, and excuses that if the Federal Republic had not supplied the facilities France or some other country would have instead.

Knowledgeable West Germans do not make such arguments to their peers, much less to other nations' nuclear experts, nor do they for a minute tolerate such false "justifications" from other countries in international political discussions. And rightly so. The reason this paper has bothered to try to clear away the preceding arguments is that they have been annoyingly distracting to the larger debate. The subject deserves much more serious and more accurately focussed treatment than it has been getting.

United States policies on nuclear exports have proven outmoded, lacking in coherence, and in the subject controversy, inadequate. Policy has simply comprised precedent and a set of almost routine approval procedures; it has never been given explicit, coherent formulation. There is only weak provision for coordination of the many parts of the government that need to be involved (this includes at least the State Department, the Arms Control and Disarmament Agency, the Congress, the Nuclear Regulatory Commission, the Energy Research and Development Administration, the Commerce Department, the Treasury Department, the Office of Management and Budget, and the National Security Council). International considerations have not been fully integrated with domestic decisions; for example, the planning for uranium enrichment capacity has failed to accommodate foreign needs along with domestic ones. Fundamentally, much of the wavering on international issues stems from ambiguities in domestic nuclear policy itself; uncertainty over the future demand for nuclear power within the U.S. has induced uncertainties about the nation's foreign enrichment commitments, as we have seen in the Brazilian case. Overall, the policies have failed to anticipate the weakening of hegemonical control that has slowly been occurring. And they have not been designed and pursued in such a way as to draw high-level attention to these difficult issues.

The past few years have revealed considerable confusion about the export of sensitive technologies. In the context of discussions over whether the enrichment enterprise should be turned over to private industry, the Energy Research and Development Administration announced that it would welcome offers for U.S. participation in a large multinational enrichment venture, without specifying what kind of venture. At about that time Assistant Secretary of State Dixy Lee Ray was actively encouraging export of nuclear technology in general, which others in the government opposed. The matter was put to a test in early 1975 when a Bechtel Power Corporation official went to Brazil and intimated that Brazil might be able to buy into the private Uranium Enrichment Associates venture that Bechtel was leading. As was later revealed in congressional testimony, Bechtel's business development manager John A. Damm wrote a letter dated 21 March 1975 to Minister of Mines and Energy Shigeaki Ueki, saying:

> There has been a most recent decision by ERDA to encourage UEA to seek potential sites for enrichment plants outside the U.S.A. One of the locations which is likely most promising is Brazil, with the abundant hydro potential in the Amazon basin.

When news of this intimation got back to Washington, Dr. Damm was caused to follow up with a second letter which included a furnished "I-have-been-asked-to-convey-the-following" paragraph noting that there seemed to have been a misunderstanding and that the United States government had no present intention of providing enrichment technology to Brazil. No matter what this episode was really all about (whether it was a "fishing expedition" or a genuine misunderstanding, or whatever), it came late in the sequence of events in Brazil and was largely irrelevant; it is not likely to occur again. But it illustrates the kind of confusion incoherent policy can lead to.

There are a number of issues on which the United States must now formulate policy. One of the most urgent of these is that of assuring the worldwide supply of enriched fuel in such a way as to keep recipients confident of supply, thus maintaining a disincentive against the acquisition of indigenous enrichment technology by countries not now possessing it. Last year's events in Brazil were a telling example: in public rhetoric, and no doubt in internal discussions, the hastily imposed moratorium on U.S. fuel promises was interpreted as a threat to Brazil's independence and stable energy development. Such doubts should not have had to arise. In cooperation with other countries the United States now has to make new enrichment plans. No matter what those plans turn out to be, whether they provide for a national or a multinational enterprise, they will lose their antiproliferatory value if they are allowed to become an unreliable bargaining currency to help the United States get its way on other matters. As long as the foreign purchasers keep their promises, fuel that is guaranteed should be delivered.

While we are on the subject of enrichment, we should mention that questions are arising, as they always have before, over whether new techniques of enrichment should be placed under security classification by their developers. The United States has to decide what to do about its laser-based processes. The West German nozzle process has been left in the public domain from the start, to which the United States has apparently never objected, so the time for its classification may have passed; however, that may not be entirely true, since substantial changes in the original designs may have to be made as the process is moved toward commercial operation. (It has been said that the nozzle process does not hold a proliferatory threat, because it cannot be used to enrich uranium to above 4 percent; but it does not seem impossible, with substantial —admittedly difficult—modification of the pipes, valves, compressors, and other ancillary equipment, to convert a nozzle enrichment facility to make highly enriched uranium for explosive devices.)

In recent years the United States has reacted to several potentially proliferatory events inconsistently and too late. It made little protest when the Indian explosion went off, for instance, but now is experiencing a rising sense of indignation and must decide whether to embargo its supply of fuel to India or take other censuring action. In the Brazilian case, parts of the American government, notably several congressmen, protested vociferously when they learned of the sale of the sensitive technologies; but there is no indication that Secretary of State Kissinger protested to the Germans (Chancellor Schmidt has publicly denied receiving any protests from the United States over the issue, and his denial has never been challenged). The sending of the four-man technical team from the United States two months after the sale was announced in the nuclear trade press, whatever its value, cannot be construed as a timely, high-level diplomatic representation. Irrespective of what particular position should have been taken on the issue, that its seriousness was not recognized earlier, and that broader discussion and higher representations were not made, indicates the rather low rating of these events as policy questions.

As we will describe shortly, several issues are arising on which the United States can take an important stand. Those include questions of the conditions for export of special nuclear material (particularly to non-NPT countries), the development of the IAEA, and the international management of the nuclear fuel cycle.

International policy issues. A question that seems to have been allowed to fade into the background, but which deserves renewed consideration, is whether the exporting nations should place their reliance totally on trilateral exporter-importer-IAEA safeguards, or whether, with especially sensitive exports at least, the strategic transfer should not be made under a bilateral safeguards pact as well. The bilateral undertaking might not have to be much different from the strictest of the nation-to-nation agreements now in force, and it might simply reinforce the statements of restraint in use of the transferred goods, and reserve for the supplier the right of occasional observation or verification. This would not need to be done for every transfer, but for diplomatically delicate situations it could lend reassurance, supplement the IAEA audits, and place the obligations of monitoring and enforcement more squarely upon the nations that have assumed the risk and reaped the benefits of the export. Although, as we have insisted several times before, IAEA safeguarding is essential, the problem with depending solely on that arrangement is that it could–in a worst case, but of course that is what we must worry about–simply be used as a scapegoat; nations should not be absolved of direct responsibility in these matters just because collective monitoring mechanisms have also been relied upon.

There is now a great burden on Brazil to demonstrate its good intentions and to assume a role of leadership in international nuclear politics. Many decisions lie ahead for Brazil: how, specifically, to carry out its construction plans, training programs, and utility-development strategies; how to manage the environmental nuisances that will be generated; how to relate to Argentina as the two countries approach the nuclear threshold; how to appraise the nuclear weapons option for the longer future; and how to fit itself into the regime of international nuclear control. Brazil is to be commended for renouncing use of the imported technology to make any nuclear explosive device and for accepting the IAEA safeguards provisions. Now, having thus endorsed in spirit the principles of the Nuclear Non-Proliferation Treaty, there would seem to be no reason for Brazil not to sign that treaty, place its entire nuclear enterprise under IAEA safeguards, encourage Argentina to do the same, ratify the Treaty of Tlatelolco, and thus weave itself into the set of solemn promises and restraints which hold the nuclear world together.

Soon the advanced nuclear nations will have to make a number of crucial decisions affecting the long-term future. One important set of decisions has to do with internationalizing and managing the nuclear fuel cycle. The supply of enriched uranium has to be assured. Techniques need to be developed and arrangements made for managing the growing quantities of nuclear waste. Major decisions have to be made about the interrelated issues of reprocessing, uranium and plutonium recycling, waste management, breeder reactors, and plutonium control. None of this can be accomplished without heavy financial and political commitment, technological innovation, and the devising of sophisticated, novel institutional arrangements. The initiative and example taken by the developed nations will be crucial.

Another set of impending decisions has to do with finding a way to manage the export competition so that nonproliferation objectives will be preserved even as the nuclear power market expands. Nothing before has prompted so much concern over this issue as the Brazilian case has; this particular deal gained high visibility because of its financial magnitude, the size of the commitment to nuclear technology made in this single step, and the possibility that northern hemispheric security was compromised simply for commercial gain. The export pressures of capital and manpower overcapacity and the drive for export revenue, in a market comprising a surprisingly few developing countries, coupled with the import incentives of energy independence, prestige, and pursuit of the weapons option, have led to a kind of competitiveness that urgently demands to be brought under control. The money market is a key factor; the Brazilian purchase illustrates the kind of risks European banks are now taking, and as one West German official remarked, "Our most tenacious competitor is not Westinghouse or General Electric, but the Export-Import Bank" (which has backed most American reactor exports).

It has been observed, in retrospect, that it would have been better if the U.S. government had pulled its vendors out of Brazil, agreeing to let the West Germans take the entire reactor sale, along with the ancillary sales, if they would refrain from including the sensitive technologies. For whatever reasons, this did not happen. And it could have happened only once. But this line of thought suggests that what is needed is a suppliers' cartel arrangement, or set of governmental agreements that would have an equivalent effect.

This would retard proliferation. Besides, it would stabilize the reactor vendors' market, making sales more predictable over the long run; for the manufacturers of these products costing hundreds of millions of dollars apiece and financed over many years, such coordination would make sound commercial sense. Hopefully the London suppliers' conference will be able to work out a solution. So far the conference has managed at least to secure agreement among the exporters to consult each other on major sales decisions.

It is worth repeating that IAEA safeguards are essential and deserving of much greater political and financial support. And in no way should they ever be compromised in reactor sales competition.

The proposals for multinational fuel facilities need to be evaluated. The case for such centers is by no means clear, yet. Although it sounds nice to say that regional facilities should be built in different parts of the world, it is in the most troubled regions that such centers are least likely to prove workable. India and Pakistan? Argentina and Brazil? The Middle East? Experience with the Urenco and Eurodiff enrichment consortia in Europe has demonstrated that multinational facilities can indeed be operated successfully, albeit with difficulty. However, their inherent disadvantages should be recognized: unless elaborate precautions are taken, the facilities will simply spread the technology, as technicians from the participant countries observe the operations; in the extreme, such plants can be expropriated by the host country; and, with reprocessing in particular, promotion of the technology by a multinational group, even on a pilot scale, tends to encourage by example its development elsewhere. There are advantages, more than economic; the partners in a venture can presumably "keep an eye on each other." The Brazilian facilities have been pointed to as being "multinational," but it should be recognized that as the plants come into operation they will increasingly become Brazilian properties, staffed by Brazilian technicians.

If the United States decides to become serious about such proposals, it should take initiative and set a firm example. Perhaps it should develop a waste-treatment center with Canada. Or, if it decides to go ahead with its Barnwell reprocessing plant, perhaps it should open that facility to international participation.

Proliferation issues should be accorded much higher political attention. Governments do not generally give proliferation problems high priority, but leave them to specialists. Other issues are allowed to override them. In many ways the oil shock was felt more strongly than the shock of the Indian explosion was. Disappointingly little international protest was made when that blast was set off, for instance. The issues being long-term, difficult, and arcane in appearance, struggling with them is not seen as a way to earn votes. Higher officials get involved only in "emergencies."

It is regrettable that in the West German-Brazilian case neither the Bundestag, the Brazilian Congress, nor the U.S. Congress became seriously involved until the decisions had virtually been sealed, and even since then the treatment has in general remained rather uninformed and desultory. The Bundestag has never held a single full discussion of the Brazilian deal. Neither has the Congress of the United States. Nor has either of those parliaments devised a comprehensive nuclear export policy. It therefore remains illusory to speak of the German position" or "the American policy."

The current situation. With nuclear technology, military and civilian capabilities overlap strongly. There are few technical ways to avoid the duality of the knowledge, materials, and facilities. The choices on the use of this dual technology are largely political ones. Traditional hegemonies have weakened. The technological barriers are slowly coming down, so it is no longer impossible for developing nations to build nuclear devices. The sources of supply of nuclear goods have diversified. It is now possible to "buy into" the nuclear club, at least part way. If the world needed a demonstration of the above points, it was given that by India in 1974.

Proliferation is inevitable. The question is at what rate, where, with what timing, and thus with what unstabilization and disjointedness. Nuclear decisions have a certain irreversibility about them, and all have long-term consequences. Renunciation of nuclear innocence leads to a progressive hazarding of the various nuclear thresholds: securing nuclear technology, exploring the idea of weapons, making a commitment to build a first explosive device, detonating that device in a test, refining the bomb design, developing delivery capability, stockpiling more sophisticated weapons, threatening to fire them in anger, and so, tragically, on.

The mere existence of knowledge in the laboratories tempts exploitation of it, "just to keep the options open." Once the options become established, pressures even from minority sources may encourage next steps. And as the nuclear weapons states have learned, once large promotional bureaucracies are spawned they develop lives of their own, as do the specialized capital-intensive laboratories and industries that supply the technologies. Further, what one country does usually induces neighboring countries to follow suit, or at least "to keep *their* options open." Around the world belligerent pairs and clusters will warily be trying to decide what moves and countermoves to make.

If proliferation is inevitable, what is the value of trying to retard it? Certainly there is no reason to make it trivial to acquire the sensitive technologies. The world can use the precious time secured to draw countries further into treatied obligations and mutual interdependencies, to elicit statements of intention, to devise technical and political means of monitoring and surveillance, and to place controls on plutonium and other materials. We still have some time, but we must move quickly to intensify the public debate and take some of these measures, because it looks as though, one way or another, nuclear futures will continue to be sold.

Chapter 15

Relevance of Intelsat Experience for Organizational Structure of Multinational Nuclear Fuel Facilities

Eugene B. Skolnikoff

The development of space communications capability by the United States in the early 1960s, and the desire to establish a global operating system, led to the creation in 1964 of the International Telecommunications Satellite consortium (commonly called Intelsat). The organization has been effective in building and operating a space communications system that serves its 86 national members, though there have been important political differences and problems along the way. The experience of that organization can offer lessons for the design of international institutional structures in other areas, especially those that are concerned with high technology subjects. In particular, there may be important lessons for a multinational nuclear fuel facility that will face at least some of the same problems.

In brief summary, Intelsat was formed in 1964 largely as a result of U.S. initiative motivated by a desire to maximize the new technology the United States developed. Membership was open to all interested countries, but the Soviet Union and other Eastern European countries, with the exception of Yugoslavia, elected not to join. The reasons for the Russian decision were varied, but among others reflected their much lower external communications requirements. Because of the voting and cost-sharing arrangements adopted, this latter factor would have resulted in a minor Soviet voice in the management and operation of the organization. (The Soviet Union and China are users, however, with ground stations in operation or under construction.)

Costs and organizational influence were apportioned according to likely system use, which gave formal expression to the U.S. technological dominance at the time. The system development and management was entrusted to Comsat, a quasi-private U.S. entity, which also provided the U.S. representation in the organization. However, the original agreements were established on an interim basis on the insistence of the Europeans, with a requirement for renegotiation after five years (not completed until 1972). A permanent organization was then created, considerably altered from that established by the earlier agreement.

In the permanent agreement, U.S. dominance was substantially reduced, and the role of Comsat was to be phased out in favor of an internal system manager. In addition, a multitiered governing structure was created designed to give each member nation equal voice over general policy questions, but to allow weighted voting on technical management and operation in favor of those countries with the technical knowledge and greatest stake in the system.

In terms of actual results, Intelsat's performance has been remarkable in achieving a high-capacity operating system in a short time. By 1969 it had provided global coverage, and by 1972, satellite utilization costs had been cut by two-thirds (Levy, 1975). By 1975 the two operational satellites over the Atlantic—the fourth hardware generation that had been developed—provided approximately 3000 active channels, with more possible. That amounted to two-thirds of available capacity, which itself had grown by more than a factor of ten in a decade (Chayes, 1975).

These achievements were not realized without controversy, nor are operational effectiveness and progress the only criteria for evaluating success. But certainly the ability to implement and manage high-technology systems is an important achievement, especially among international organizations and a *sine qua non* for the prospects for a multinational fuel center. Other issues that were encountered in the Itelsat experience are also relevant to the design of such a center, along with an understanding of the basis of Intelsat's technical performance; these can be more easily developed in a framework relevant to the multinational fuel center by focusing first on the important requirements that must be met in its design.

IMPORTANT INSTITUTIONAL REQUIREMENTS

In laying out the more important requirements for a successful multinational fuel center, it will be assumed that it could be involved in any or all of the technical tasks of enrichment, reprocessing, storage of spent fuel, fuel fabrication, and long-term waste storage. In the list that follows, no priority of importance is implied, nor can the requirements be thought of as being mutually exclusive. They obviously interact in important ways, and are even at times in conflict. In addition, some "requirements" are actually issues for decision, rather than requirements in a strict sense.

1. *Operational Capability.* Clearly, a multinational fuel center must have the capability to carry out the functions assigned to it, which can cover a wide range including operation of facilities, transportation, provision of physical

protection, adequate monitoring of nuclear material, construction, and other functions.

2. *Efficiency.* The enterprise should be seen as a cost minimizer, especially in comparison with national facilities, and be able to conduct its business with acceptable dispatch and cost.
3. *Physical Security.* The facility must be able to preserve physical security and control all nuclear material in its charge—especially with respect to its host country.
4. *Stability, Guarantees, Reliability.* Long-term stability must be assured as a disincentive for the development of national facilities. Similarly, fulfillment of commitments must be able to be guaranteed at least as reliably as those of national facilities.
5. *Shared Decision-Making.* All countries relying on the center must have a voice, and believe they have a voice, in the governance of the facility. There can be no substantial dependence on the internal political process of another sponsoring state, which also implies clear independence or insulation from the host country. The decision and allocation processes must also be seen to be fair and nondiscriminatory or to favor one group of countries (poor nations, for example) only in agreed ways.
6. *Confidence.* The governance of the enterprise must be such as to build confidence among member states, which implies, among other attributes, openness of operations and high quality of staff.
7. *Innovation, R&D.* The multinational fuel center must be able to take advantage of new technology and avoid freezing technological designs in what may prove to be uneconomic configurations. This may imply an R&D capability, or at least a structure that allows application of R&D results. Also, the existence of the organization should not become a barrier to national R&D.
8. *Access to Technological Information.* A prime rationale for a multinational center is to discourage proliferation by deterring the construction of dangerous national facilities. It could have the effect of accelerating proliferation by giving access to technology to countries unable to obtain it easily other ways. The operation must attempt to minimize this possibility.

LESSONS OF INTELSAT

In the light of these general requirements, what lessons can be drawn from the Intelsat or other experience that bears on their realization? A one-to-one correspondence is not possible, of course, but many potentially useful observations can be made. The remarks below will be organized in the framework of the requirements presented in the previous section; inevitably some seemingly arbitrary categorization will result, and in some cases the Intelsat experience will offer little guidance.

Operational Capability. Probably the most significant aspect of the Intelsat experience with regard to technical performance was the early dominance of the country—the United States—with the greatest technical capability, the major investment, and the greatest incentive to achieve an operating system. There were essentially no technological alternatives at the start, so that U.S. technology effectively had a monopoly. It is interesting to note, however, that the Europeans were unwilling to accept the new technology (they had veto power since a communications link requires two ends) without at least some voice in its deployment. The limited duration of the Interim Agreement also guaranteed them an opportunity eventually to revise the relationship.

The effective U.S. monopoly was also exercised in the use of Comsat as system manager. In effect, this action established a relatively apolitical organization in the key technical slot and served thereby to enhance system technical efficiency (though it also cost the U.S. government some measure of political control).

The situation of technological monopoly does not apply in the case of nuclear facilities, though the relevant technologies are not widespread. However, it can be expected that giving major technical voice, especially at the outset, to those countries with the clear technical capability is more likely to achieve efficient technical operation. Efficient technical operation is of course not the sole objective of a multinational fuel center, so that some broader form of political control is also necessary (see below).

The use of a quasi-private system manager, such as a "Comsat," is probably an inappropriate model for the nuclear situation which has more widely shared technology than in the communications satellite case. However, the idea of clearly separating ownership from technical management, which worked so well for Intelsat, is an important generalizable proposition. Whether the manager be a private company or government consortium is a secondary matter in this respect.

Some other lessons in the area of operational capability are also relevant:

- Representation on the management body of individuals from the relevant technical agencies of member governments is a further step to emphasize technical qualifications in the operation of the facility.
- The area of responsibility of the enterprise should be clearly spelled out and limited. One reason for the relative success of Intelsat, and of other international organizations with operating responsibility, is the strict construction of its functions and sphere of operation. At times this can lead to the denial of a useful additional role in a new field for a capable organization already in existence. But the confusion of purposes and the likely conflicts that result outweigh the benefits.
- The founding countries should be those with a clear self-interest in the success

of the organization. Progress is likely to be rapid, as in Intelsat, and possible disagreements minimized when seeking a common goal.

–National quotas on personnel should be avoided in the management body. It is desirable to have some representation from all member governments if possible as a way of achieving participation and demonstrating fairness. But formal national quotas can quickly destroy, or lead to loss of confidence in, the technical competence of an organization.

Efficiency. Efficiency, interpreted as efficiency of operation, entails the same comments as above with regard to operational capability.

There is another aspect of efficiency, however, which is a result of achieving economies of scale. In the early days of Intelsat, the economies that could be realized by developing and deploying large-capacity satellites and creating a global monopoly seemed clear. An original U.S. objective, in fact, was to spell out in the basic agreement a requirement that all signatories must use Intelsat facilities rather than develop any alternatives of their own. That proved infeasible, and presumably unnecessary economically since other specialized systems outside Intelsat are appearing.

In the case of a multinational fuel center, however, the idea of monopoly is essential, and thus economies of scale must be achievable as a means of deterring national facilities. By contrast with Intelsat, highly restrictive covenants may have to be built into the original agreement, though that raises questions of discrimination between states that already have their own facilities and those that would be precluded from building their own.

Physical Security. Intelsat experience seems to have little bearing on this requirement, other than competent management.

Stability, Guarantees, Reliability. Intelsat experience has some, perhaps quite significant, lessons to offer about this requirement that is such an important one in the functioning of a multinational fuel center.

As new communications capability is deployed, it rapidly becomes an essential aspect of a nation's economy. For many of the member countries, there is no real alternative to Intelsat for external communications. Yet nations have been willing to rely on the system in full knowledge that it would be technically possible suddenly to be denied access. They have thus been willing to depend on the operation of an international organization they do not control for an essential service.

Some of the characteristics of the mechanism for governance, discussed below, affect this level of confidence. Presumably the sense of participation in the policy process is an important facet. Some of it may depend as well on the technical nature of the issues from which contentious political questions have been largely abstracted by limiting further expansion of Intelsat into new areas. Thus, Intelsat remains largely in the background and is not seen as a political bargaining chip. Furthermore, the system is essential in some measure to all participants so that it is not evident who would gain by disruptions of service; thus there can be high confidence that there is little likelihood of denial of service.

But the fact remains that countries are relying on an international organization for the operation of a technological service essential to them.

Shared Decision-Making and Confidence. Some of the most significant lessons lie in this area. The Intelsat structure arrived at in the permanent arrangements is an interesting model that offers the hope of providing both political control and operational efficiency.

Essentially, Intelsat now has a three-tiered structure. First, there is a universal membership body operating on a one-nation/one-vote basis. Actually, it is two bodies, one with representation of foreign affairs ministries, and the other with representation of telecommunications ministries. The former deals with broad general policy issues, and the latter with more general budgetary and investment concerns.

A board of governors exercises direct oversight of the system, with weighted voting that favors the United States and other large users who have the major stake and investment in the system. However, wide geographical representation is assured, and no single nation can dominate the system.

Finally, taking over from Comsat, there will be an international administrative and technical staff, under a director-general, to manage the system under the direction of the board.

The arrangement is designed to satisfy the needs and demands of small countries for a voice in the governance of the system so they can feel able to protect their interests without destroying the effectiveness of management. Some such arrangement would seem to be essential for a multinational fuel center.

There are reasons other than simply a *sense* of participation why such a structure would seem advisable. One is the danger that the organization could operate as a cartel with regard to those dependent on it, so that some form of representation for all involved seems necessary in price-setting, scheduling, and related matters.

Another is that there are more economic interests involved than simply the nuclear material (or the communications capacity). In particular, there are important development and construction contracts to be let, with significant economic and technological implications. Intelsat was originally following the practice, through the decisions of Comsat, of giving contracts primarily to firms that had the technological advantage, which tended to mean U.S. firms. The Europeans strenuously objected and were able to get some favorable treatment for a time. However, the permanent arrangements, under the pressure of the developing countries, returned to a policy of contracting based on technological quality and least cost. The same kind of issues could arise in the operation of a multinational fuel center. Similarly, any policies to give advantages to poorer countries will require, or be seen to require, their participation in policymaking.

Finally, the issues raised earlier about security of supply, guarantees, and stability all require a structure of governance that cannot easily be dominated by one or a few countries to the disadvantage of others, and that develops confidence and trust in its continuity and fairness.

It is also worth noting that there is no need to go through the occasionally painful "interim" period of Intelsat. That period of heavy U.S. dominance aroused much antagonism and often led to disputes in which the United States frequently had to retreat. Given the Intelsat experience, it ought to be possible, in many areas at least, to start with the "Definitive Arrangements" rather than with an interim structure.

Innovation, R&D. Intelsat's record in stimulating R&D and accepting innovation in general has been good, as witnessed by the rapid evolution of the system. That capability is obviously related to the technical qualifications of the manager, Comsat. It is also related to the contracting process which in effect relies on private firms as the source of innovation. Those firms have the competitive incentives—both contracts from Intelsat and contract opportunities from other industries—to maintain technological leadership. On the other hand, there has been dissatisfaction in what some countries perceive as premature freezing of technological options, for example, in setting ground station designs that led to excessive investment in large ground stations and now obsolescent facilities.

This general area will be particularly important for a multinational fuel center since new technological developments (such as new enrichment technology) could defeat the primary purpose of deterring national facilities. There seems no simple answer to this problem from the Intelsat experience except high technical quality, attention to the problem, and avoidance of premature technological decisions.

Access to Technological Information. The very requirement for openness and participation as a necessary aspect of governance and confidence enhances the possibility that a multinational facility will actually help to spread nuclear technology instead of deter proliferation. In this area, Intelsat experience may be a poor guide, for there was no parallel concern about proliferation. Obviously some kind of a balance will have to be struck in the detailed agreements that attempt to minimize transfer of technology, while satisfying equally important considerations of participation and equity.

U.S. Representation. An added set of issues peculiar to the United States can be gleaned from the Intelsat experience. The U.S.-chosen instrument for satellite communications, Comsat, created a series of special problems that ought not be repeated for a multinational fuel center. The quasiprivate character of Comsat created two kinds (at least) of problems:

1. It made collaboration with government entities of other countries unnecessarily awkward, at least.
2. It circumscribed the ability of the U.S. government to maintain political control of its representation in the system.

Whatever justification there may have been in the communications case to create a special instrument of that kind, it would seem particularly unwise in the nuclear field where the issues strike so close to the heart of central political interests. In that connection, it is of concern that present U.S. policy is aimed at moving enrichment technology into the private sector.

CONCLUSION

The multinational fuel center will be a much more difficult development than creating an international organization for communications satellites, but there are obvious parallels that can be used or avoided. The relative success and effectiveness of Intelsat gives hope that the task can be achieved.

REFERENCES

Chayes, A. "U.S. Policy-Making for Satellite Communications," Appendix B of *The Committee on the Organization of the Government for the Conduct of Foreign Policy*. Washington: U.S. Government Printing Office, June 1975.

Levy, S.A. "Intelsat: Technology, Politics and the Transformation of a Regime," *International Organization* 29 (Summer 1975).

Part IV

Political Setting

Chapter 16

International Arrangements for Nuclear-Fuel-Cycle Facilities: The Politics of the Problem

William Epstein

THE PROBLEM IN PERSPECTIVE

It is worth recalling that the Three-Power Declaration made by President Harry S. Truman, Prime Minister Clement Attlee, and Prime Minister W.L. Mackenzie King on November 5, 1945 was intended "to prevent the use of atomic energy for destructive purposes [and] to promote . . . the utilization of atomic energy for peaceful and humanitarian ends." The declaration stated that "no system of safeguards that can be devised will of itself provide an effective guarantee against production of atomic weapons by a nation bent on aggression. . . . We are, however, prepared to share, on a reciprocal basis with others of the United Nations, detailed information concerning the practical industrial application of atomic energy *just as soon as effective enforceable safeguards against its use for destructive purposes can be devised*" (emphasis added).

It is somewhat surprising that 30 years later, not only these three countries but also France, West Germany, the Soviet Union, and others are engaged in selling and transferring nuclear materials, equipment and technology that could facilitate the proliferation of nuclear weapons, without "effective enforceable safeguards."

While reactors for power production are large and expensive undertakings that take five or more years for construction and start-up, research reactors of a few megawatts in size are relatively simple and easy to build. Moreover, once any relatively advanced country has a power or research reactor, it is not difficult for it to build a small reactor by itself. Such reactors can be utilized or "programmed" to produce the maximum amount of plutonium-239 if their operation for generating power at the lowest possible cost is not the object.

A small research experimental reactor using natural uranium as a fuel (which produces somewhat more plutonium than one using enriched uranium) can be built at relatively low cost. And since low-grade uranium resources are present in many countries, such reactors can be used to produce plutonium in sufficient quantities to make one or two bombs a year from indigenous resources.

As far back as 1966, Dr. Glenn Seaborg, then chairman of the U.S. Atomic Energy Commission, stated with regard to the process of separating weapons-grade plutonium from the fuel irradiated in civil power reactors: "It is perfectly feasible to build a clandestine chemical processing plant using readily available technology and equipment." Ten years later, it is even more feasible to do so. It has been estimated that a separation plant capable of producing 15 to 20 kilograms of plutonium-239 a year (enough for two or three explosive devices) could be built in a year or two by any reasonably advanced country at a cost of $1-3 million. Even if the cost has been grossly underestimated, it is clear that the amount of money involved is not large.

If French and West German companies proceed to sell plutonium reprocessing plants, even though under international safeguards, to any country not a party to the Treaty on the Non-Proliferation of Nuclear Weapons (NPT) that promises not to use the plutonium to explode a nuclear device for either military or peaceful purposes, the danger of the spread of nuclear weapons will be greatly increased, because of the possibility that these promises may be disregarded at some later time.

Among the near- and potential-nuclear powers that are not parties to the NPT—India, Pakistan, Israel, Spain, South Africa, Argentine, and Brazil—all but South Africa have, or are reputed to have, or have arranged to acquire, some type of reprocessing plant. South Africa, having a uranium-enrichment process, has no need for a plutonium reprocessing facility if it should decide to manufacture a nuclear weapon.

There is at the present time no actual commercial need to reprocess spent-oxide fuel from power reactors into plutonium. In fact it is not at all clear when a need will arise for plutonium either as a fuel for commercial power reactors or for peaceful nuclear explosions. For the next 25 years it seems likely that uranium and enriched uranium can meet all requirements. Nor are the economic advantages of plutonium recycling and breeder reactors very clear in themselves or in the light of the potentialities of uranium enrichment by some laser or other new process or of developing some alternative energy source.

It is difficult to resist speculating as to the reasons that motivate countries now to acquire or retain their own national reprocessing capabilities. It seems reasonable to assume that at least some of these countries wish to have their own reprocessing facilities in order to acquire a real nuclear-weapons option.

Even before the Indian nuclear explosion in May 1974 there was a very marked erosion of faith in the NPT. Many non-nuclear countries, in particular the have-not countries, had become or claimed to have become disillusioned by the failure of the nuclear powers to live up to their obligations under the NPT. The nuclear powers have not halted tests of nuclear weapons or the nuclear arms race (nor have they agreed to any measures of nuclear disarmament), although these were clearly contemplated by the NPT. In fact, many countries regard the SALT agreements as mere blueprints for the continuation of the nuclear arms race by the two superpowers under agreed rules and conditions.

The nuclear powers have likewise done very little to fulfill their obligations under the NPT (Article IV) to facilitate the peaceful uses of nuclear energy, "especially in the territories of non-nuclear-weapon states party to the Treaty, with due consideration for the needs of the developing areas of the world." And although the potential benefits of peaceful nuclear explosions are increasingly dubious, Article V of the NPT provides a specific way for dealing with this issue. It stipulates that "negotiations on this subject shall commence as soon as possible after the Treaty enters into force." Nevertheless no negotiations have taken place except perhaps in the context of the Threshold Test Ban Treaty of July 1974. That treaty and the follow-on treaty signed on May 28, 1976, permitting the U.S. and the U.S.S.R. to conduct both military and peaceful nuclear explosions up to 150 kilotons in yield, have spread further doubts about the sincerity of the two superpowers.

The disappointment and frustrations of most of the developing countries at what they regard as breaches of the treaty provisions by the nuclear powers have weakened the NPT. Its ratification on the eve of the NPT Review Conference by five Euratom countries (West Germany, Italy, the Netherlands, Belgium, and Luxembourg) and by South Korea in April 1975 and by Japan in May 1976, helped to preserve some of the vitality of the treaty. Most Third-World countries, however, do not regard these accessions as important since they concern mainly developed countries that are allies of the United States and are therefore under the American nuclear umbrella.

In the ultimate analysis, whether a country decides to "go nuclear" and acquire nuclear explosives depends on how it views its needs and interests in the context of the military, political, economic, and moral climate of the world. A country will evaluate its position on this issue not only by what the nuclear powers do or do not do to honor their obligations under the NPT, but, more importantly, by what it regards as threats or potential threats to its security from its neighbors or other countries in its region of the world. Thus, Pakistan is obviously much more concerned by its fear of what India may or may not do than it is about what the nuclear powers may or may not do in living up to their obligations under the NPT. This applies to Israel and the Arab states, South Africa and the other African states, Argentina and Brazil, and others. Public opinion in these countries will respond more to the actions of neighboring countries than to policies of the nuclear powers or nuclear supplier countries.

What these and other great powers do, however, determines not merely the military situation, but also the political and moral climate of the world. Their actions can be used as arguments or pretexts by either the doves or hawks in any country. Thus the failure of the nuclear powers to abide by their obligations under the NPT fosters a more acceptable climate and provides a more facile excuse for any country to acquire nuclear weapons, or at least obtain or retain that nuclear option.

Apart from the failure of the two superpowers to implement their commitments, we have witnessed the negotiation of a sale of a reprocessing plant by France to Pakistan and of a complete nuclear fuel cycle (including uranium-enrichment and plutonium reprocessing facilities) by West Germany to Brazil, albeit under international safeguards. In addition, the announcement in March 1976 that Canada had negotiated an agreement to resume nuclear aid to India was greeted with incredulity and dismay by many people. In view of the strong stand taken by Canada in condemning the Indian nuclear explosion and in immediately cutting off all further nuclear assistance to India, it was argued that the resumption of nuclear cooperation even under safeguards would represent a form of Canadian condonation of the Indian explosion. Fortunately, after a sharp debate in Parliament and further study of the matter, Canada reversed itself. On May 18, 1976 the Canadian government announced that it had decided against any further nuclear cooperation with India, since India had refused to agree that no Canadian supplies of nuclear technology, equipment, or materials—past or future—be used to manufacture any nuclear explosive device, and that all nuclear facilities involving Canadian technology be placed under safeguards. This Canadian decision not to resume nuclear aid to India was one of the more positive developments in an otherwise disturbing scene.

It is in the light of these developments and circumstances that we must view and assess the credibility and viability of any proposed international arrangements for nuclear-fuel-cycle facilities as a measure to curb the proliferation of nuclear weapons.

HOW EFFECTIVE AND ENFORCEABLE ARE INTERNATIONAL SAFEGUARDS?

The primary purpose of the IAEA is to promote and facilitate the peaceful uses of atomic energy for the welfare of mankind. International safeguards were intended not as an end in themselves but only as a means to ensure that the nuclear assistance provided by or through the IAEA would not be misused for military purposes.

The IAEA Statute and its safeguards system do not prevent or prohibit states from acquiring nuclear weapons or material and the equipment to make them. France and India, for example, were members in good standing when they first exploded nuclear devices, as are the United States, the Soviet Union, and the United Kingdom, all of which have continued to conduct explosions.

The Statute and safeguard system contained in the agency's document INFCIRC /66, adopted in 1965, are intended to ensure only that the specific projects, facilities, and nuclear equipment which a country has agreed to place under IAEA safeguards are not diverted to military purposes. A second, and separate, safeguards system adopted in 1971 and contained in INFCIRC/153 applies to non-nuclear-weapons states that are parties to the NPT and that have accepted the obligation not to acquire nuclear weapons. The agency's safeguards under the NPT, unlike the 1965 safeguards under the IAEA Statute, are designed to apply to *all* nuclear materials in *all* peaceful nuclear activities within a state (after mining the uranium ore), as required by Article III(1) of the NPT, to ensure that such material is not diverted to nuclear weapons or other nuclear explosive devices.

Neither of the IAEA safeguards systems can prevent diversion of fissionable material; at best they can only serve to detect any diversion sometime after it has occurred, and thus have only a deterrent effect. Any nation, however, that is interested in doing so can find ways to delay or hinder the effective operation of the safeguards system.

IAEA safeguards do not provide for any physical control of nuclear material. The IAEA/NPT safeguards system is a system for keeping track of nuclear material primarily on the basis of auditing the accounts of a national government or of its nuclear facilities, supplemented by automatic instrumental checks and by periodic inspections. In some ways the process is similar to that of the auditors of a bank, but the IAEA inspectors have much less authority than do bank auditors, and their task would be much more difficult. In a bank, the flow of dollars and securities in and out of the system can be more readily checked. (And even under those circumstances, banks make mistakes or are victims of theft by outsiders or embezzlement by insiders.)

The IAEA/NPT safeguards system can only be as effective as the accounting and physical security safeguard techniques of each state permit and as that state is willing to cooperate with the agency. If a state regards the safeguards as contrary to its national interest at any time, it is very doubtful that it would enforce or accept them (for example, international surveillance or inspections), even if this amounted to a unilateral abrogation of its agreement. If any state should decide at any time that it wished to divert nuclear materials to the manufacture of weapons or nuclear explosive devices it would not be very difficult for it to circumvent, suspend, or nullify the safeguards, without going to the bother of actually falsifying its records.

There are no real measures of sanction or enforcement in case of noncompliance or violation by any state. The only recourse is under the provisions of Article XII(C) of the IAEA Statute. This provides merely that the Board of Governors can give publicity to the violation by reporting it to the members of the Agency and to the Security Council and General Assembly of the United Nations. In addition, the Board may direct the curtailment or suspension of any assistance being provided by the Agency, and the Agency may suspend any noncomplying member from membership.

It should be noted that there is nothing in the NPT or the IAEA safeguards systems to prevent or make illegal the possession of weapon-grade uranium or plutonium. It is only their diversion to weapons or explosive devices by non-nuclear states that is banned.

In any case, a state wishing to free itself from its commitments need not engage in any clandestine or prohibited activities or even go into the black market to acquire nuclear materials and equipment. It can, if it chooses to do so, openly withdraw from the NPT on three-months' notice and be free to do what it wishes. It may risk the suspension or termination of further nuclear aid (as India did), but it is not easy to visualize even such sanctions as the withholding of food and other non-nuclear aid or the suspension of trade occurring, let alone more drastic measures.

As indicated, the IAEA/NPT safeguards system is a system that seeks to deter diversion only by the risk of early detection. The continuing usefulness of such a system of safeguards depends on the moral climate and on the strength of the entire nonproliferation regime. In the final analysis, even domestic law, which is much more effective since it can be enforced by a local police power and a system of penalties, depends for its observance on the prevailing climate of opinion and on the respect in which the laws are held. The same applies, to a large extent, to the NPT and its safeguards. The respect accorded the treaty, and the observance of its obligations by the parties to it, may have more to do with the effectiveness of its safeguards system than the actual measures and procedures themselves. Unfortunately, many of the provisions of the NPT have been ignored rather than honored.

RESTRICTIONS ON EXPORTS AND THE LONDON "SUPPLIERS' CLUB"

As a result of its unfortunate experience with India, Canada unilaterally announced the imposition of what it describes as the strictest export regulations of any nuclear supplier country. On December 20, 1974, Canada announced that it would in the future require that international safeguards to ensure against the development of any nuclear explosive device cover: (1) all nuclear facilities and equipment supplied by Canada; (2) all nuclear facilities and equipment using Canadian-supplied technology; (3) all nuclear material—uranium, thorium, plutonium, and heavy water—supplied by Canada, and all fissionable material produced from or with these materials; and (4) all nuclear materials, whatever their origin, produced or processed in facilities supplied by Canada.

But even these strict regulations can be evaded. A country that is not a party to the NPT, which has obtained a research or power reactor using natural uranium, can build its own and can operate it with its own natural uranium, since most countries have their own resources of uranium ore or access to foreign sources.

If it buys or builds a small reprocessing plant (as India did), it can acquire its own stockpile of plutonium. Thus a country subject to the Canadian safeguards and abiding strictly by them can still manufacture its own nuclear explosive devices. Only the NPT requires that *all* nuclear material in *all* peaceful nuclear activities of a country be placed under international safeguards. Anything short of that can only delay or inhibit a country from going nuclear.

Unless export regulations require that all of a recipient's nuclear materials and activities be put under IAEA safeguards, they permit the continuation of the invidious discriminatory situation wherein countries that are not parties to the NPT can obtain nuclear materials, equipment, and technology on more favorable terms (that is, under less strict conditions) than parties to the treaty.

Thus the Canadian export regulations, and those of other countries, even if they are as strict as the Canadian, are not good enough to prevent a country from acquiring a nuclear explosive device. And if they are not good enough, they are inadequate.

In all fairness, it must be said that Canada and several other countries have urged in the IAEA, at the U.N. and other international bodies, and in the London Suppliers' Club, that the nuclear exporting countries should go all the way and refuse to supply nuclear materials, equipment, and technology unless the recipient country agrees to put all its nuclear material and facilities under IAEA safeguards. Mainly because of the widely reported opposition of France and West Germany, the other members of the Suppliers' Club (the U.S., U.S.S.R., U.K., Canada, and Japan) have not been able to agree on such strict export regulations.

While some progress has been made, the measures that have been agreed are only partial ones and are not sufficient to prevent the proliferation of nuclear weapons.

The meetings of the Suppliers' Club are conducted in the utmost secrecy (a procedure not conducive to great or early progress) and even the provisions of the consensus thus far achieved have not been made public and are not fully known. But enough has been announced or leaked out to make it clear that they are still inadequate. The sale of reprocessing plants is not barred and safeguards do not extend to the full nuclear program of the recipient countries, including indigenous sources, nor do they cover all aspects of nuclear technology.

There are serious grounds for doubting even whether safeguards over the entire nuclear fuel cycle would by themselves be adequate to prevent the further spread of nuclear weapons. But one would have thought that insistence on such broad safeguards would be the minimum required; such unfortunately is not the case.

Although the export regulations imposed by Canada, France, West Germany, and other suppliers provide that the nuclear *technology* transferred cannot be used for any purpose other than that agreed in the agreement of sale—which presumably means that the facilities and equipment supplied cannot be copied by the purchaser—there is no practical way to enforce the restriction. Even allowing the presumption of copying specified in the West Germany-Brazil agreements, the remedy or recourse provided is only to cancel future nuclear assistance.

While the seven members of the Suppliers' Club have been joined by eight other countries namely, Belgium, Czechoslovakia, East Germany, Italy, Poland, the Netherlands, Sweden and Switzerland, several supplier or potential supplier countries—such as India, South Africa, Australia, Argentina, Brazil, and China (although China has not up until now been a supplier or given any indication of becoming one)—are not members.

Loopholes still exist and the danger of further nuclear proliferation will remain unless all supplier and potential supplier countries join the club, and even then the club would have to adopt adequate measures to ensure that additional countries do not obtain the means to acquire a nuclear explosive capability. Dr. Sigvard Eklund, Director-General of the IAEA, stated at the United Nations on May 13, 1976:

> It is my conviction therefore that the further spread of nuclear explosive technology cannot be prevented by the rules laid down so far. Unless we go further, we shall sooner or later see the addition of new nuclear weapon states. . . ."

NUCLEAR-FUEL-CYCLE CENTERS

Various proposals have been put forward for the establishment of regional, multinational, or international fuel-cycle centers for uranium enrichment, fuel fabrication, reprocessing, spent-fuel storage, and waste management. While such centers might help to prevent the acquisition of fissionable material for explosive devices by any individual national government, the problems involved in their establishment and operation are considerable. The economic, financial and administrative problems that would have to be solved are outside of the scope of this paper. Assuming they were all capable of solution, however, the overriding political problems might still prove to be insuperable.

The very reasons that motivated a country to remain outside of the NPT would probably also be decisive in keeping it from participating in any international and perhaps even binational venture. Apart from any loss in closing down its own national plant, and irrespective of the economic benefits it might gain from participating in a large-scale multinational project, these financial considerations would be relatively insignificant for any state that believed it was important for its national security to go nuclear or to retain the option to do so.

The establishment of nuclear-fuel-cycle centers might be a useful undertaking as one of a number of measures to help contain the risks of nuclear proliferation, but it would not alone add a great deal to the present nonproliferation regime.

ADDITIONAL MEASURES FOR EFFECTIVE SAFEGUARDS

If the nuclear supplier states are really serious about trying to prevent the further proliferation of nuclear weapons, there are several measures that would enable them to do so. Given basic agreement on the goal, a program of measures that would minimize the risks of further proliferation to as near zero as is possible in a world of sovereign states is simple to prepare. The measures themselves are also technically simple and relatively easy to supervise and control. Only the political will is needed to adopt and implement them.

Such a program to be effective and adequate should include the following measures:

1. The nuclear-weapons states that are parties to the NPT (U.S., U.S.S.R., and U.K.) should agree to accept IAEA safeguards over all their peaceful nuclear activities. This would remove the discriminatory features that apply only to the non-nuclear parties to the NPT, which the latter resent.
2. Since the nuclear parties already have more than enough weapons-grade U-235 and Pu-239 for any conceivable military use, there should logically be no obstacle to their halting the production of any further fissionable material for military purposes and agreeing to place the entire fuel cycle of their plants and facilities under safeguards.
3. The nuclear-weapons states and supplier states should agree to make no sales and provide no nuclear material or equipment to any nuclear or non-nuclear country unless that country agrees either to become a party to the NPT or to accept equivalent restraints and to place *all* of its nuclear material and facilities, including re-exports, under IAEA safeguards.
4. The supplier states should make a public declaration (and not conceal it by a consensus declaration in the London Club) and have it adopted by the IAEA Board of Governors and General Conference, and by the U.N. General Assembly, that peaceful nuclear explosions under national control are incompatible with the peaceful utilization of nuclear energy.
5. All spent fuel should be returned to the supplier country for storing, or sent to some other place where it can be stored safely until such time as reprocessing is required.
6. All spent fuel should be reprocessed only at such time as reprocessing is required and becomes commercially feasible, and preferably at some internationally operated and safeguarded plant from which it would be exported

only as fuel for identified peaceful purposes. If reprocessing is done at cost, or better still is subsidised, it would provide a substantial inducement to countries not to acquire or use their own reprocessing plants.

7. Nuclear supplier countries should not export any chemical-separation plant for plutonium reprocessing that might come under national control. Nor should they export any nuclear material, equipment, or technology to any non-nuclear country that has or that acquires any reprocessing plant under national control. Those countries that already have reprocessing plants of any kind should be required to place them under international management and operation.
8. The supplier countries should work toward the adoption of physical security systems, both national and international, which would prevent the theft of nuclear material by terrorists and criminal elements, either at nuclear facilities or during transport of nuclear materials.
9. The nuclear supplier countries should adopt effective sanctions providing for the immediate withholding of *all* nuclear assistance by *all* suppliers not only for future contracts but also under previous or existing contracts, from countries that violate any of the above measures.

David E. Lilienthal, former chairman of the Tennessee Valley Authority and the first chairman of the U.S. Atomic Energy Commission, stated in evidence to the Government Operations Committee of the U.S. Senate in January 1976:

> The tragic fact is that the atomic arms race is today proceeding at a more furious and more insane pace than ever. Proliferation of capabilities to produce nuclear weapons of mass destruction is reaching terrifying proportions. We have to decide now what we can do, now, within our own capabilities, to prevent a very bad situation from becoming a disastrous and inevitable one. I therefore propose as a private citizen that this committee with its great prestige, call upon the Congress and the President to order a complete embargo to the export of all nuclear devices and all nuclear material, that it be done now, and done unilaterally. Further, unilaterally, the United States should without delay proceed by lawful means to revoke existing American licenses and put an end to the future or pending licensing to foreign firms and governments of American know-how and facilities paid for and created by American taxpayers' funds and American brains.

Dr. Hans Bethe, a Nobel Laureate who contributed to the production of the first A-bomb, and Dr. Herbert York, director of the Lawrence Radiation Laboratory at Livermore, California during the development of the H-bomb, also testified and supported the idea of an embargo on all nuclear exports until a treaty is concluded "that really assures control over proliferation."

In March 1976 the chairman of the Senate Committee on Government

Operations, Senator Abraham Ribicoff, proposed that the United States enter into a cooperative arrangement with the other suppliers, including France and West Germany, that would guarantee each supplier a minimum number of reactor sales a year, in a sort-of-cartel arrangement, to eliminate harmful cut-throat competition. If agreement on strict export controls and market-sharing arrangements could not be reached, he proposed that the United States should unilaterally stop supplying reactor fuel to France, West Germany, and any other countries that failed to meet the American requirements. Domestic pressures might be generated in those countries that would cause their governments to change their policies.

Such actions are obviously very drastic and very serious in their import and could not be lightly undertaken. Their mere public ventilation and discussion, however, might produce the desired result. In such matters, open publicity rather than secrecy would seem to be the best course.

All countries, including advanced countries such as France and West Germany, and supplier or potential states such as South Africa and India, neither of which is a party to the NPT, ought to regard the prevention of the further spread of nuclear weapons as in their interest. Hence it is possible that even nonparties to the NPT might be willing to support and subscribe to measures, such as those listed above, that are not dependent on or related to adherence to the NPT.

While the above suggestions for technical measures to strengthen the safeguards system are far-reaching in nature, they are no more radical than is the concept of nonproliferation itself or, for that matter, the Non-Proliferation Treaty. If the nuclear-weapons states and the other supplier states would agree to accept and implement them, the non-nuclear states might be persuaded to do likewise. And if the two superpowers were to lead the way in this regard, they could help to create a momentum and community of interest in favor of the adoption of such a strengthened and expanded safeguards system similar to that which they succeeded in achieving when they worked out the provisions of the Non-Proliferation Treaty and obtained the commendation of the United Nations by an overwhelming majority.

While the foregoing list of safeguards measures would reduce the possibilities of nuclear-weapons proliferation to the maximum degree that is technically possible by such means of control, they would not and could not ensure that all further proliferation was stopped. As the three heads of government foresaw in their prescient Declaration in 1945, no system of safeguards could of itself provide an effective guarantee against the production of atomic weapons by a nation bent on acquiring them.

POLITICAL MEASURES

As previously indicated, what may be of even greater importance in preventing the spread of nuclear weapons is the removal of the incentives that motivate countries to acquire these instruments of destruction. That will require a series of "political" measures and accords that are even more salient than the "technical" safeguards outlined above. These political steps would be aimed at what I would call the "psychological denuclearization" of nations and peoples—that is, a rational reduction and elimination of the desire or need for nuclear weapons in the interests of a more secure world.

The most important political measures are the following:

1. A declaration by all nuclear powers not to use or threaten to use nuclear weapons against any non-nuclear state that does not have any nuclear weapons in its territory. China and France have already announced their support for such a declaration, and in 1966 the Soviet Union submitted a similar proposal under the "Kosygin Formula." The United States and United Kingdom have also given a somewhat similar pledge to the countries that are members of the Latin American Nuclear Free Zone under the Treaty of Tlatelolco. Hence it should not be beyond the ingenuity of diplomats to work out some agreed formula for such a declaration. If each of the nuclear powers also undertook to give some positive security assurances that it would, subject to the U.N. Charter obligations, come to the aid of any such state attacked or threatened with nuclear weapons, this would help to strengthen the feelings of confidence and security of non-nuclear states.
2. A declaration of non-first-use of nuclear weapons by each of the nuclear powers. This would at a single stroke reduce nuclear weapons to a similar legal status as that of chemical and biological weapons under the Geneva Protocol.
3. An immediate moratorium by the United States and the Soviet Union suspending all underground nuclear tests for five years and the commencement of negotiations to achieve agreement on a permanent ban of such tests.
4. A joint declaration by the United States and the Soviet Union that they would undertake to conduct underground nuclear explosions for peaceful purposes, either for themselves or for non-nuclear states, only when authorized to do so by some competent international authority to be established for that purpose, and that they would begin immediate consultations to convene a conference for the creation of a nondiscriminating and equitable special international regime for peaceful nuclear explosions, as required by Article V of the NPT.
5. A declaration by the two superpowers of a qualitative freeze on the production of new nuclear weapon systems for a fixed period of years pending the negotiation of a treaty to that effect. This declaration would entail the suspension of all flight testing of new delivery vehicles and weapons systems.
6. The immediate commencement of negotiations at SALT and elsewhere for concrete visible progress toward substantial reductions of existing nuclear weapons systems, far below the inflated levels set by the Vladivostok accord.

The above list of political measures to help achieve psychological denuclearization and remove or counterbalance the incentives for nuclear proliferation is not exhaustive, but is illustrative of the kinds of political actions that would create the right climate for nonproliferation in a more secure world.

Only a combination of political, psychological, and technical measures can succeed in providing the required credibility and viability for an effective nonproliferation regime.*

REFERENCES

Eklund, Sigvard, Dr., Director-General IAEA, Statement at United Nations, New York, May 13, 1976.

Epstein, William, *The Last Chance: Nuclear Proliferation and Arms Control.* New York: The Free Press, 1976.

IAEA Documents: INFCIRC/66 and revisions thereto, and INFCIRC/153.

Ikle, Fred C., Dr., Director, U.S. Arms Control and Disarmament Agency, Statement: "Illusions and Realities about International Approaches to Nuclear Energy", Washington, D.C., May 13, 1976.

Macdonald, Flora, M.P. Canada, Address in Montreal, Canada, May 20, 1976.

NPT, *Final Document and Records of the Review Conference of the Parties to the Treaty on the Non-Proliferation of Nuclear Weapons.* Geneva, 1975.

Parliamentary Debates (Canada), March 23 and May 18, 1976.

Ribicoff, Abraham, Senator, Article on Op-Ed Page, *New York Times*, 28 March, 1976.

Sanders, Benjamin, *Safeguards Against Nuclear Proliferation.* SIPRI, Stockholm: 1975.

Seaborg, Glenn, 1966.

Senate, United States, 1975, *Peaceful Nuclear Exports and Weapons Proliferation.*

Senate, *United States Congressional Record*, 1975/76.

SIPRI, 1974, *Nuclear Proliferation Problems.*

SIPRI Yearbooks, 1969–1976, *World Armaments and Disarmament.*

*After the presentation of this paper, the Canadian Government progressively imposed stricter regulations on nuclear exports. In November 1976 it announced that it would halt all further nuclear assistance to Pakistan if the latter proceeded with its purchase of a plutonium reprocessing plant from France. On December 22, 1976, it announced that nuclear "shipments to non-nuclear-weapon states under future contracts will be restricted to those which ratify the NPT or otherwise accept international safeguards on their entire nuclear program." This new policy extended the previous coverage to "what a country received from other suppliers or what it might do on its own." Thus Canada is the first country in the world that has adopted a nuclear export policy that goes all the way with respect to the scope of coverage by safeguards.

Index

About the Contributors

C. Allday, Managing Director, British Nuclear Fuels Limited, Member of the Board of Urenco Ltd., Member of the Beirat of Centec GmbH, Member of the Beirat of United Reprocessors, GmbH.

C. Beets, Director R&D Safeguards, Centre d'Etude, de l'Energie Nucleaire, Mol, Belgium.

Albert Carnesale, Associate Director of the Program for Science and International Affairs, Harvard University.

Abram Chayes, Felix Frankfurter Professor of Law, Harvard Law School.

Peter Dyne, Director, Office of Energy Research and Development, Department of Energy, Mines and Resources, Ottawa, Ontario, formerly Director, Chemistry and Material Science Division, Whiteshell Nuclear Research Establishment, Pinawa, Manitoba (Co-Authors, R.B. Lyon and D.R. McLean, Chemical Technology Branch, W.N.R.E., Pinawa, Manitoba.)

William Epstein, Special Fellow of the United Nations Institute for Training and Research and a Visiting Professor at the University of Victoria, Victoria, British Columbia, Canada. Formerly Director of the Disarmament Division of the United Nations, and from time to time Special Consultant on Disarmament to the Secretary-General of the United Nations. Latest book, *The Last Chance: Nuclear Proliferation and Arms Control*.

F.R. Farmer, Safety Adviser to the Authority, UK Atomic Energy Authority, Culcheth, Warrington, England; Visiting Professor, Imperial College of Science and Technology, London.

D.A.V. Fischer, Assistant Director General for External Relations of the International Atomic Energy Agency.

Ted Greenwood, Assistant Professor, Department of Political Science and Center for International Studies, Massachusetts Institute of Technology; Research Associate, Program for Science and International Affairs, Harvard University.

S.R. Hatcher, Director, Applied Science Division, Atomic Energy of Canada Limited, Whiteshell Nuclear Research Establishment, Pinawa, Manitoba, Canada. (Co-Author, W.W. Morgan, Chemical Technology Branch, W.N.R.E., Pinawa, Manitoba.)

W. Bennett Lewis, Distinguished Professor of Science, Physics Department, Queen's University, Kingston, Ontario; formerly Senior Vice-President, Science, Atomic Energy of Canada, Limited.

William W. Lowrance, Research Fellow, Program for Science and International Affairs, Harvard University.

George W. Rathjens, Professor of Political Science, Massachusetts Institute of Technology.

Lawrence Scheinman, Senior Policy Analyst for International Policy Planning, USERDA. (On leave from Cornell University.)

Mark Sharefkin, Senior Research Associate, Resources for the Future, Inc., Washington, D.C.

Eugene B. Skolnikoff, Director, Center for International Studies, and Professor, Department of Political Science, Massachusetts Institute of Technology.

Constance Begley Smith, J.D. Candidate, Harvard Law School.

Mason Willrich, Director of the Division of International Affairs of the Rockefeller Foundation, and John C. Stennis Professor of Law, University of Virginia (on leave).